The ultimate fate of
the universe

T0296963

The ultimate fate of the universe

JAMAL N. ISLAM
Reader in Mathematics, The City University, London

CAMBRIDGE UNIVERSITY PRESS
CAMBRIDGE
LONDON · NEW YORK · NEW ROCHELLE
MELBOURNE · SYDNEY

CAMBRIDGE UNIVERSITY PRESS
Cambridge, New York, Melbourne, Madrid, Cape Town, Singapore, São Paulo, Delhi

Cambridge University Press
The Edinburgh Building, Cambridge CB2 8RU, UK

Published in the United States of America by Cambridge University Press, New York

www.cambridge.org
Information on this title: www.cambridge.org/9780521113120

© Cambridge University Press 1983

This publication is in copyright. Subject to statutory exception
and to the provisions of relevant collective licensing agreements,
no reproduction of any part may take place without the written
permission of Cambridge University Press.

First published 1983
This digitally printed version 2009

A catalogue record for this publication is available from the British Library

Library of Congress Catalogue Card Number: 82-14558

ISBN 978-0-521-24814-3 hardback
ISBN 978-0-521-11312-0 paperback

Contents

Preface

In 1977 I wrote a short technical paper entitled 'Possible ultimate fate of the universe' which was published in the *Quarterly Journal of the Royal Astronomical Society*. A number of colleagues found this paper amusing. Just then, Weinberg's excellent book *The first three minutes* appeared and it occurred to me that it would be interesting to have a book about the end of the universe. Soon I was requested by the astronomical magazine *Sky and Telescope* to write a popular version of my paper for them. This appeared in January 1979 under the title 'The ultimate fate of the universe'. The response to this article convinced me that a popular book on the subject would not be inappropriate. The result is this present book.

I have written the book with the person who has no special scientific knowledge in mind. All the technical terms mentioned and all the physical processes described are explained in as simple language as I have been able to use. However, I have avoided oversimplification. This means that some parts of the book will require close attention by the reader who does not have any scientific background, but I hope that everyone who cares to read the book will be able to follow the main ideas without much difficulty.

I have made free use of some of the books and articles mentioned in the bibliography for the more standard parts of this book. As this material is very standard, I feel it unnecessary to acknowledge the sources individually. However, I have tried, wherever possible, to mention the names of the people who have been responsible for originating new ideas or making new observations. I have usually given full names and year of birth and death of past scientists who are reasonably well known. With some exceptions this information has been taken from the 1970 edition of the *Encyclopaedia Britannica*. I have found the latter useful for some other pieces of information. For contemporary scientists I have used initials instead of first names.

I am deeply grateful to F.J. Dyson for encouraging me to write this book in the first place and for the fact that many of the new ideas in this book are his. My thanks are also due to S.J. Aarseth, S.W. Hawking, S. Mitton, J.V. Narlikar, M.J. Rees and J.C. Taylor for useful comments on various aspects of this book, and to the staff of Cambridge University Press, in particular Marion Jowett, for their cooperation. I am deeply indebted to Mrs Mary Wraith for her efficient typing of the manuscript. Lastly, I would like to thank my wife Suraiya and my daughters Sadaf and Nargis for constant support and encouragement during the period in which this book was written.

November 1982 JAMAL N. ISLAM

To the memory of my parents.
Also, to the memory of my brother
Tareque Muinul Islam (1930–1979)
and to the memory of my nephew
Nadeem Omar (1955–1979)

Note on some conventions

In this book the term 'billion' is used in the American sense to mean a thousand million. The number represented by 1 followed by n zeros is often written as 10^n. Thus a billion is 10^9 and a billion billion is 10^{18}. The reciprocal of 10^n, that is, 1 divided by 10^n, is written as 10^{-n}. Thus a billionth is 10^{-9} and a billion billionth is 10^{-18}. Also, 10^{10^h} is the number represented by 1 followed by 10^n zeros.

1

Introduction

What will eventually happen to the universe? The question must have occurred in one form or another to speculative minds since time immemorial. The question may take the form of asking what is the ultimate fate of the Earth and of mankind. It is only in the last two or three decades that enough progress has been achieved in astronomy and cosmology (the study of the universe as a whole) for one to be able to give at least plausible answers to this kind of question. In this book I shall try to provide an answer on the basis of the present state of knowledge.

To appreciate the possibilities for the long-term future of the universe it is necessary to understand something of the present structure of the universe and how the universe came to be in its present state. This will be explained in some detail in Chapter 3. In this introduction, I shall briefly outline the contents of this book to provide a 'bird's eye view' to the reader. All the terms and processes mentioned in this summary will be explained in more detail in the succeeding chapters.

The basic constituents of the universe, when considering its large-scale structure, can be taken to be galaxies (Fig. 1.1), which are 'islands' of stars with the 'sea' of emptiness in between, a typical galaxy being a congregation of about a hundred billion (10^{11}) stars (e.g. the Sun) which are bound together by their mutual gravitational attraction. The galaxy that we inhabit (together with the Sun and the system of planets of the Sun, called the solar system) is referred to as the Milky Way or simply the Galaxy. The universe can be defined as the totality of all galaxies which are observable and others which

Fig. 1.1. A rich cluster of galaxies in the constellation Fornax, showing a variety of structural types. The cluster is held together by the mutual gravitational attractions of its member galaxies. In 10^{27} years, a large cluster such as this may be reduced to a single black hole smaller than the smallest galaxy shown.

are causally related to the observable ones. There are strong indications that, on the average, galaxies are spread uniformly throughout the universe at any given time.

It is found observationally that all galaxies are receding from each other so that the universe is not static but is in a dynamic state. This recession of the galaxies from each other is referred to as the expansion of the universe. From the rate at which galaxies are moving away from each other it can be deduced that all galaxies must have been very closely packed about 10–20 billion years ago. It is generally believed that at that time there was a universal explosion in which matter was thrown asunder violently. Later the matter condensed into clumps, to become the galaxies of the present time. The recession of the galaxies is a remnant of the initial explosion, the so-called 'big bang'.

One of the most important questions in cosmology – to which the answer is not definitely known – is whether the expansion of the universe will continue forever, or whether the expansion will halt at some time in the future and contraction set in. The model of the universe which expands forever is usually referred to as the 'open' universe, while that which stops expanding and begins to contract is called the 'closed' universe. Thus one of the most pressing questions in cosmology is whether we live in an open or a closed universe. The ultimate fate of the universe depends on the answer to this question. There are some indications that the universe is open, but this is by no means settled.

What will happen to the universe eventually if it is open? Since the basic constituents of the universe are galaxies, we can examine this question by asking what will happen in the long run to a typical galaxy in an open universe. Consider, then, a typical galaxy. It consists mainly of stars. All stars evolve with time and eventually die, that is, they reach a final stage after which very little further evolution takes place, at least in time scales of tens of billions of years. There are three such final stages for a star, namely those of white dwarf, neutron star and black hole. These final states will be explained in detail in Chapters 6 and 7. For the present it will be sufficient to note that these are states in which matter is in a highly condensed form, the most condensed being a black hole. Given sufficient time, all stars in the galaxy will die, that is, reach their final states of white dwarf, neutron star or black hole. We refer to stars in these three states as dead stars. Sufficient time in this case is between a hundred and a thousand billion years or perhaps longer. Thus, in about a thousand billion years the galaxy will consist of dead stars and cold interstellar matter in the form of planets, asteroids and smaller pieces of matter, still bound together in their mutual gravitational attraction. Different galaxies will of course continue to recede from each other, so that the average distance between galaxies will be much longer than at present.

The next significant changes in the galaxy will take place over a much longer time scale, in which a substantial number of the dead stars will be ejected from the galaxy altogether by coming into close collisions with other stars. In a billion billion (10^{18}) or a billion billion billion (10^{27}) years or so, 99% of the dead stars may be ejected from the galaxy in this manner. The remaining 1% of the dead stars will form a very dense core which will eventually coalesce into a single black hole whose mass will be about a billion solar masses. We can call this the 'galactic black hole'. The process described in this paragraph will be referred to as the stage of dynamical evolution of the galaxy.

I have defined the three final states of dead stars as states in which very little further change takes place in time scales of tens of billions of years. When time scales very much longer than billions of years are considered, these final stages do change. In fact a black hole of the mass of the Sun does radiate in very minute amounts and thus continues to lose its mass. A black hole of solar mass will disappear altogether by this radiation process in about 10^{65} years, which is very much longer than the time a galaxy takes to reduce to a single black hole. This radiation of a black hole is not significant while the dynamical evolution of the galaxy proceeds. However, once the galactic black hole has been formed, one can ask whether this will last forever or whether it will suffer further changes. In fact the galactic black hole will evaporate completely in about 10^{90} years. A supergalactic black hole, that is, one formed out of the collapse of a large cluster of galaxies, will evaporate completely in about 10^{100} years. Thus in 10^{100} years or so all black holes will disappear and all galaxies in the universe will have been completely dissolved. The universe will then consist of stray neutron stars and white dwarfs and other smaller pieces of matter that were ejected from galaxies during their dynamical evolution. These dead stars and pieces of matter will be wandering singly in the ever-growing and vast emptiness.

There will be some slow and subtle changes in the remaining pieces of matter over time scales which are long compared with 10^{100} years. What will be the ultimate form of the remaining pieces of matter? Here we come to the crucial question of the long-term stability of matter, the answer to which is not known. Some possibilities will be discussed in Chapters 10 and 14. One possibility is that white dwarfs and neutron stars will collapse spontaneously into black holes and subsequently evaporate, as suggested by the laws of quantum mechanics. The time scale for this is $10^{10^{76}}$ years! (If I write the word 'billion' a billion times, the resulting number will be minute in comparison with $10^{10^{76}}$.)

What about the long-term survival of civilization and of life in an open universe? It is almost impossible to predict what forms living organisms will take in the long run assuming they can survive. However, the survival of civilization and of life depends on the availability of a source of energy, and one can discuss the latter. It will be seen in Chapter 11 that, at least in principle, there will be adequate sources of energy available for 10^{100} years or so. Beyond this time civilization will have to face the problem of surviving indefinitely on a fixed finite amount of energy. This is an unresolved question but some of the possibilities will be considered in Chapter 11.

The picture presented above is likely to prevail if the universe is open. What if the universe is closed? Suppose the universe turns out to be closed in such a manner that when it reaches its maximum expansion, the average intergalactic distance is about twice that of the present time. Then this maximum will be reached in about 40 or 50 billion years. After reaching this maximum it will be almost as if a movie film of the universe were taken until the time of maximum expansion and then run backwards. After about 90–110 billion years, the universe will become very dense and hot and soon afterwards there will be the so-called 'big crunch', in which all matter will be engulfed in a fiery implosion. There is very little chance of survival of any form of life in this case. What happens after the big crunch, or whether there is an 'after', is not known.

I should emphasize that the picture presented in this book is on the basis of the present state of knowledge. Even this proviso must be further qualified. The basis of this book is a model of the universe known as the standard model, which will be explained in detail in Chapter 3. I think it is fair to say that a substantial majority of cosmologists believe that the standard model is correct in its essentials. However, there is a small minority of cosmologists which adheres to the concept of one or other of some non-standard models. We shall not be concerned with the non-standard models in this book, with the exception of one, the steady state theory, which will be discussed briefly in Chapter 13. The reader will also notice that the black hole features prominently in this book. A black hole has not yet been discovered, although there are powerful theoretical and some indirect observational reasons for believing in the existence of black holes. There may be respectable scientists who do not believe in black holes, but it would appear that a majority of experts in gravitational theory subscribes to the view that black holes must exist. In this book we shall assume that black holes do exist.

The picture presented above of the open universe changes somewhat if we consider the possibility, which has recently been put forward by some physicists, that the proton, which is a constituent of all matter, is unstable with a long life-time. That is, it is conjectured by some physicists, for reasons which will be explained in detail in Chapter 14, that all protons will eventually disintegrate. This possibility has important bearing on the far future of the universe and we shall discuss this in Chapter 14.

Why should one bother about the ultimate fate of the universe? One answer to this question is similar to the answer to the question about climbing Mount Everest: because the problem exists. It is in the nature of the human mind to seek incessantly new frontiers of knowledge to explore. The ultimate fate of the universe and of civilization is an interesting problem, not least because, as we shall see in the course of this book, it raises fundamental questions in physics, astronomy, biology and other branches of knowledge, the answers to which, if they can be found, may lead to important advances in these fields.

2

Our Galaxy

In astronomy one uses distances and periods of time large compared to terrestrial ones. The word 'astronomical' has in the English language come to mean some very large quantity. When discussing the universe as a whole one uses even larger distances and periods of time than those used in ordinary astronomy. The convenient unit for measuring distances in astronomy is not the kilometer or the mile, but the light year, which is the distance traversed in a year by light moving at the speed of about 300 000 kilometers a second (km/s); a light year is approximately 9×10^{12} km or 9 million million km. To have some idea about the light year, let us consider some familiar distances and convert these to 'light travel time'. The circumference of the Earth is about 40 000 km, so in one second light can travel round the Earth more than seven times. The distance to the Moon is 371 000 km, so it takes light between 1 and 1.5 seconds to travel from the Earth to the Moon. The mean distance of the Earth from the Sun is approximately 150 million km. This distance is covered by light in 8–8.5 minutes. The mean distance from the Sun to Pluto, the furthest planet in the solar system, is approximately 5900 million km, which distance is covered by light in about 5.5 hours. A light year is thus almost 1600 times the distance from the Sun to Pluto.

When measuring distances in the solar system, the light year is too long so astronomers also use as a unit the mean distance of the Earth from the Sun. This unit is referred to as the astronomical unit. The distance from the Sun to Pluto is about 39.5 astronomical units. One light year consists of about 60 000 astronomical units.

Astronomers often use another unit instead of the light year, namely the 'parsec' which is approximately 3.26 light years. This unit comes about as follows. As the Earth revolves around the Sun, some of the nearest stars trace out ellipses in the sky against the background of very distant stars whose directions do not change. The maximum angular radius of such an ellipse, when expressed in seconds of arc, is known as the trigonometric parallax or simply the parallax of the star. It can be shown that the reciprocal of this parallax measured in seconds gives the distance of the star in parsecs. Thus a star at a distance of 1 parsec has a parallax of 1 second of arc, and a star at a distance of 2 parsecs has a parallax of 0.5 seconds of arc, and so on. This is one method by which the distances to the nearest stars are calculated. Thus Alpha Centauri, which is the nearest star, has a parallax of 0.75 seconds of arc, so its distance in parsecs is the reciprocal of this number, that is, about 1.33 parsecs. This is equivalent to about 4.34 light years. A million parsecs is referred to as a megaparsec.

On a clear, moonless night one can see thousands of stars and also the bright, cloudy patch of light stretching across the sky, noticed since ancient times and known as the Milky Way. The stars that one sees through the naked eye and even those that one sees through an ordinary telescope, belong, together with the Sun and the solar system, to the system of stars which constitutes our galaxy. This galaxy is known variously as the Milky Way, Milky Way Galaxy, our galaxy, or simply the Galaxy. In fact the word 'galaxy' is derived from the Greek *galaxias kyklos*, meaning the milky way. I shall usually refer to it as the Galaxy. The Galaxy is in the shape of a flat disc, with the Sun and the solar system about two-thirds of the way from the centre to the circumference of the disc. When one looks in the plane of the Galaxy one sees many more stars than when one looks away from this plane. The many stars in the plane of the Galaxy appear in the sky as the Milky Way. The disc that the Galaxy comprises is about 80 000 light years in diameter and about 6000 light years thick. There is also a spherical halo of stars around the disc about 100 000 light years in diameter. The density of stars in the spherical halo is much less than the density in the disc. It seems to have been the English instrument-maker Thomas Wright who first suggested in 1750, in a book entitled *Original theory or new hypothesis of the universe*, that the Milky Way consists of stars that lie in a flat slab, a 'grindstone' extending to large distances in the plane of the slab.

With an ordinary telescope one can see many faint and cloudy patches in the sky in addition to the stars and the Milky Way. As early as 1781, the French astronomer and comet hunter Charles Messier (1730–1817) published a catalogue of 103 such objects to help other comet hunters to avoid mistaking these objects as early stages of a comet. Even today astronomers refer to the objects which appeared in Messier's catalogue by the prefix M followed by a number denoting the position of the object in the original catalogue. Thus, for example, the Crab Nebula, which is the remnant of an exploding star (more about this later) is called M1, since it was the first object in Messier's list.

Messier's list of objects, which were called 'nebulae', was added to by the German-born English astronomer William Herschel (1738–1822) and by his son John Herschel (1792–1871). William Herschel, who was originally a musician, discovered the planet Uranus in 1781 and was responsible for important advances in astronomy of the period. William Herschel made a list of about 2000 new nebulae. John Herschel continued his father's programme and in 1864 published *The General Catalogue of Nebulae* which was a list of 5079 faint objects. John Louis Dreyer (1852–1926), the Danish astronomer, improved on John Herschel's list by publishing in 1888 (with supplements in 1895 and 1908) the *New General Catalogue of Nebulae and Clusters of Stars*. This list, which was still a standard work in the late 1950s, contains nearly 15 000 nebulae and star clusters. This was a remarkable achievement considering that the observations were carried out visually with the aid of a telescope, but without the use of photographic equipment.

Many of the objects in Messier's catalogue are in fact objects within our Galaxy. His catalogue contained many 'star clusters' which are groups of stars which give a cloudy appearance because of the great number of stars in them. There are two kinds of star clusters, the first being the open cluster containing a few hundred stars loosely grouped together, such as the open cluster M67 (the 67th object in Messier's catalogue) located in the constellation Cancer. Secondly, there are the globular clusters, which are spherical distributions of about 100 000 densely-grouped stars such as the globular cluster M5 located in the constellation of Serpens. These two objects are also called NGC2682 and NGC5904, respectively, because of their position in Dreyer's *New General Catalogue*. In addition to star clusters, Messier's catalogue contained nebulae which are genuine nebulae, meaning that they are indeed clouds of dust and gas. In some places the Galaxy consists of clouds of gas and dust which contain some young stars and in which the process of star formation is taking place. Such a nebula is the Orion nebula (called M42 or NGC1976). This nebula is just barely visible to the naked eye in Orion's sword.

The Galaxy is approximately in the shape of a spiral as shown in Fig. 2.1. The whole spiral is rotating in its own plane around the centre. In the inner portion of the galaxy, near the nucleus (the central bulge), where the density of matter in the Galaxy is highest, it rotates more or less like a rigid body. In the outer part of the disc, containing the Sun and the majority of observable stars, the angular velocity diminishes outwards from the centre. That is, stars which are further away from the centre lag behind those that are nearer, somewhat like the planets in the solar system. The Sun completes its orbit around the galactic centre in about 200 million years at a speed of about 250 km/s.

The centre of the Galaxy is obscured by dust and cloud, so it is difficult to make optical observations of the centre. Much information about the structure of the inner parts has been gained by radio observations. Most of the matter in the Galaxy is in the form of stars of which there are various types. The total number of stars in the Galaxy, as mentioned earlier, is approximately a hundred billion. In addition to stars there is interstellar matter in the form of gas clouds of various kinds.

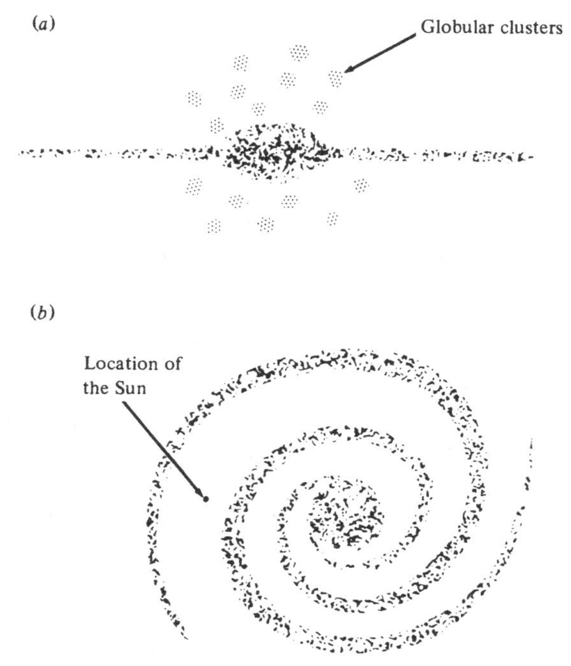

Fig. 2.1. View of the Galaxy from a point outside (*a*) in the plane of the Galaxy, (*b*) away from the plane of the Galaxy.

There are interesting inorganic and organic compounds present in the interstellar matter. The precise composition of this interstellar matter is an area of active research. There is also a magnetic field present in the Galaxy. The precise structure and description of the Galaxy is quite complicated with many unsolved problems. However, from the point of view of the large-scale structure of the universe, the detailed structure of the Galaxy is not very important.

3

The large-scale structure of the universe

Other galaxies

Many of the objects in Messier's catalogue have turned out to be systems outside our Galaxy. One of these is the Andromeda nebula (Fig. 3.1), visible to the naked eye on a clear night as a hazy patch in the constellation Andromeda. In AD 964 the Persian astronomer Abdurrahman Al-Sufi mentioned it in his *Book of the fixed stars*, calling it 'a little cloud'. The Andromeda nebula has turned out to be a spiral galaxy somewhat like our own, and a close neighbour of our Galaxy. In the late nineteenth and early twentieth centuries there was a great controversy about the nature of the nebulae listed by Messier, the Herschels and Dreyer. There was one school of thought which held the view that some of these nebulae were extragalactic, i.e. systems outside our Galaxy. In fact the original suggestion that some nebulae might be extragalactic seems to have been made by the German philosopher Immanuel Kant (1724–1804). Taking up Wright's theory of the Milky Way, in 1755 in his *Universal natural history and theory of the heavens*, he suggested that some nebulae are in fact circular discs somewhat like our Galaxy, and they are faint because they are so far away.

The controversy was finally settled in the 1920s and 1930s mainly by the American astronomer Edwin Powell Hubble (1889–1953) who demonstrated beyond reasonable doubt that most of the nebulae are indeed extragalactic. He did this mainly by measuring the so-called red shift of the nebulae, which I will explain later. After the completion of the 100-inch telescope at

Fig. 3.1. The central region of the Andromeda galaxy.

Mount Wilson near Los Angeles, in 1923 Hubble was for the first time able to resolve the Andromeda nebula into separate stars. He found a spiral structure in this nebula but the spiral structure of our own Galaxy had not been established at the time. In the spiral arms of the Andromeda nebula he found some variable stars, that is, stars whose brightness changes regularly with a certain period. Such stars were already familiar from our own Galaxy. These were known as Cepheid variables after a particular member of this class known as Delta Cephei. If one draws a graph of the brightness of these stars against time, it looks somewhat like that in Fig. 3.2. Earlier, two American astronomers, Henrietta Swan Leavitt and Harlow Shapley, had found a relationship between the observed periods of variation of the Cepheids and their intrinsic brightness. By intrinsic brightness one means how bright the object actually is, and not how bright (or faint) it appears to be to us. Technically intrinsic brightness is referred to as 'absolute luminosity' which is the total amount of light radiated by an

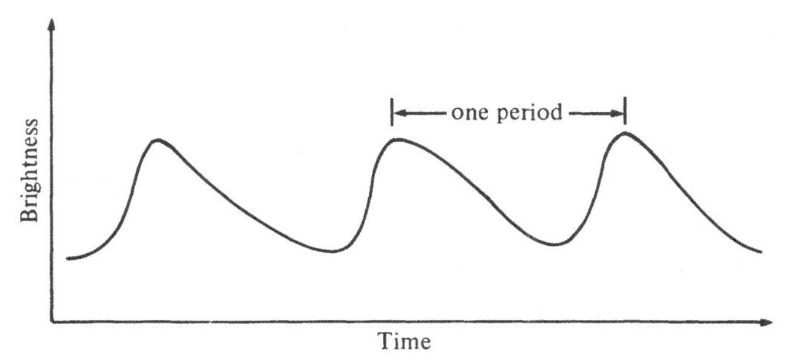

Fig. 3.2. A Cepheid is a variable star whose brightness changes in the manner shown; rapid brightening followed by gradual dimming.

astronomical object in all directions. How bright an object appears to us is measured by what is known technically as 'apparent luminosity' which is the amount of light received by us in each unit of area of our telescope. The relation between period and absolute luminosity that Leavitt and Shapley found can be shown approximately by a graph such as the one in Fig. 3.3. From the graph it is clear that from a knowledge of the period of a Cepheid variable one can deduce its absolute luminosity. Thus from the periods of the Cepheids that Hubble found in the Andromeda nebula he was able to deduce their absolute luminosity using the Leavitt–Shapley relationship.

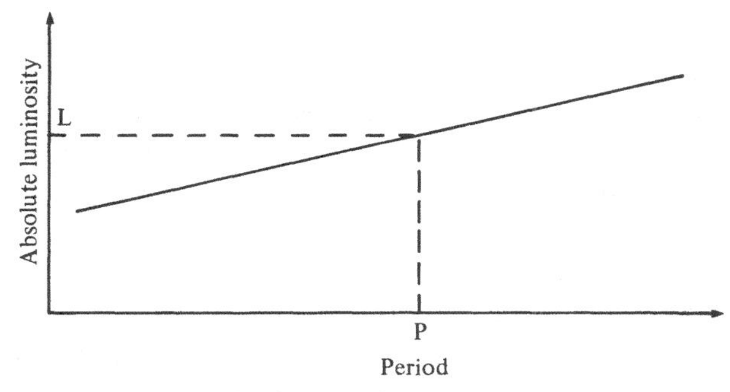

Fig. 3.3. The graph of the luminosity versus period for a Cepheid is given by the solid line. Thus if the period is *P*, the luminosity of the star is *L*.

Now if one knows the absolute and apparent luminosities of an astronomical object, one can deduce its distance, because apparent luminosity depends on the distance – the further away the object, the fainter it is. Hubble's conclusion was that the Andromeda nebula was at a distance of 900 000 light years, which was clearly outside our Galaxy, since it was more than ten times further than the most distant object known in our Galaxy. In fact in the late 1940s and early 1950s the German-born American astronomer Walter Baade (1893–1950) and others showed that there are in fact two types of Cepheid variables and that those that Leavitt observed and those that Hubble observed belong to different types, so that Hubble used the wrong period–luminosity relation. The distance to Andromeda nebula turns out to be over two million light years.

Red shift, the universe and its expansion

Hubble, with the aid of observations done on red shifts by the American astronomer M.L. Humason, established beyond reasonable doubt that many of the faint cloudy patches seen in the sky are themselves galaxies consisting of billions of stars which look faint because they are very far away. By now it is clear that as far as one can detect through the most powerful optical and radio telescopes the universe is filled with galaxies which are separated by empty or nearly empty space. Before proceeding further we should perhaps explain what we mean by the term 'universe'. We find galaxies as far as our most powerful telescopes can detect. It would not be unreasonable to assume that there are more galaxies beyond the furthest ones we can detect. Thus one way to define the universe would be to consider it to be the totality of all galaxies which are causally connected to the galaxies that we observe. We are assuming here that if there were intelligent beings inhabiting the furthest-known galaxy, they would see a distribution of galaxies around them similar to ours, and the furthest galaxy in their field of vision in the opposite direction to us would have a similar distribution of galaxies around it, and so on. The totality of galaxies connected in this manner could be defined as the universe. This raises the question, are there galaxies which are not connected to us in this manner? This question is related to an alternative definition of the universe as 'everything that exists'. The two definitions are not necessarily the same, although they may be. We prefer to use the first definition because the second raises questions (for example, 'is it possible for other universes than our own to exist?') the answers to which lie at present in the realm of pure speculation.

Galaxies tend to occur in groups called clusters from a few to a few thousand in a cluster. There is some evidence of existence of clusters of clusters, but no evidence of clusters of clusters of clusters or higher hierarchies. Observations indicate that on the average galaxies are spread uniformly throughout the universe at any given time. This means that if we consider a portion of the universe which is large compared to the distance between typical nearest galaxies, then the number of galaxies in that portion is roughly the same as the number in another portion with the same volume at any given time. Thus, since the average nearest galaxies are about a million light years apart, the number of galaxies in a cube of a hundred million light years is roughly the same no matter where the cube is situated, provided it is considered at the same time. This proviso 'at any given time' about the uniform distribution of galaxies is important because, as we shall see, the universe is in a dynamic state and so the number of galaxies contained in any given volume of the universe may change with time. Also, this proviso is an extrapolation from observation because light from distant galaxies started millions of years ago and so we have no information about these distant galaxies at the present time. The distribution of galaxies also appears to be isotropic about us, that is, it is the same in all directions from us. If we make the assumption that we do not occupy a special position amongst the galaxies, we conclude that the distribution of galaxies is isotropic about any galaxy at any given time. In fact it can be shown that if the distribution of galaxies is isotropic about any galaxy, then it is necessarily true that the galaxies are spread uniformly throughout the universe.

Hubble discovered round about 1930 that the distant

galaxies are moving away from us. He also found that this motion of the distant galaxies is systematic in the sense that the further away a galaxy is from us, the higher is its velocity of recession. This velocity of recession of distant galaxies follows a rule called Hubble's Law which states that the velocity is found by multiplying the distance of a galaxy by a certain number known as Hubble's constant (it is constant in the sense that it is the same for all galaxies at any given time). Another way of saying this is that the velocity is proportional to the distance. Thus if a certain distant galaxy has a certain velocity away from us, then a second galaxy which is twice as far away as the first one will have twice the velocity away from us than the first galaxy. This rule is approximate because it does not hold for galaxies which are very near nor for those which are very far, for the following reason. In addition to the systematic motion of recession every galaxy has a component of random motion. For nearby galaxies this random motion may be comparable to the systematic motion of recession and so nearby galaxies do not obey Hubble's Law. An obvious example of this violation is the nearby galaxy in Andromeda mentioned earlier, which has a velocity towards our galaxy instead of away from it. The very distant galaxies also show departures from Hubble's Law because for one thing light from the very distant galaxies started billions of years ago and the systematic motion of the galaxies in those epochs may have been significantly different from that of the present epoch. In fact by studying the departure from Hubble's Law of the very distant galaxies one can get useful information about the overall structure of the universe, as we shall see. For distant galaxies there is also the problem that if Hubble's Law held for indefinitely large distances, then the velocity of those very distant galaxies would also become indefinitely large. But we are told that Einstein's Special Theory of Relativity says nothing can travel faster than light. The resolution of this problem is subtle and we shall consider it later after explaining about red shifts.

How did Hubble find out the velocities of distant galaxies? He used the so-called red shift of the light emitted from distant galaxies. This can be understood as follows. If an observer is standing on the side of a road and if a car approaches with a siren sounding, then as the car passes the observer the pitch of the siren goes down. The higher the speed of the car, the higher will be the difference in siren pitch between the approaching and receding car. If one knows the frequency of the siren as heard in the car, it is a matter of simple calculation to find out the speed of the car by comparing the initial frequency with the frequency received by the observer. If a source is giving out signals such as sound or light, then the initial signal corresponds to a certain number of vibrations per second. When an observer receives these signals, the number of vibrations he receives depends on his speed with respect to the source. If the source is approaching him the vibrations crowd together and he receives a higher number, whereas if the source is receding the vibrations spread out and he receives a smaller number. This principle was discovered by the Austrian physicist Christian Johann Doppler (1803–1853) and is known as the Doppler effect. The Doppler effect for sound was tested by the Dutch meteorologist Christoph Hendrik Didericus Buys-Ballot (1817–1890) with an orchestra of trumpeters in an open train.

Hubble used the Doppler effect to determine the velocity of distant galaxies. He did this by comparing the frequency of light received by us with the frequency of the same light as emitted by the distant galaxies. How did he find these out? This can be understood as follows. Light is a form of electromagnetic radiation; a typical element of this radiation can be pictured as a wave like that shown in Fig. 3.4. (This picture should not be taken too literally.) The distance between

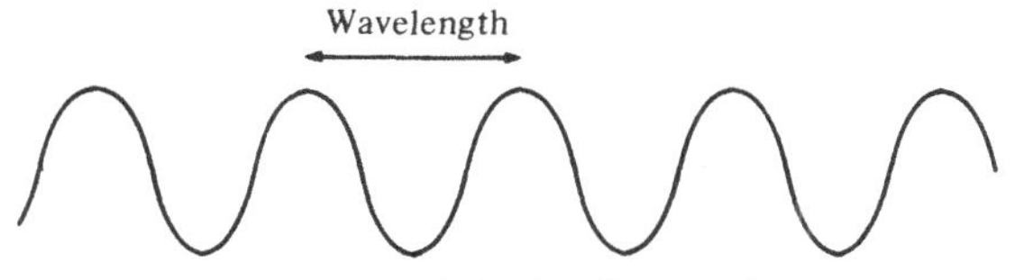

Fig. 3.4. The wavelength is the distance between successive crests of a wave.

successive crests is called the wavelength. When the radiation has wavelengths from about 0.00002 to 0.0001 cm, we call it 'light' because our eyes are sensitive to wavelengths in this region. Thus light is a form of electromagnetic radiation with wavelengths in a certain range. Radiations with wavelengths larger than those of light correspond successively to infra-red radiation (heat), microwaves and radio waves. Radiations with shorter wavelengths than those of light correspond successively to ultra-violet rays, X-rays and gamma rays. The approximate wavelengths corresponding to these kinds of radiation are given in Table 3.1. Electromagnetic waves of all wavelengths travel with the same velocity, the velocity of light. The frequency of the radiation is defined as the number of waves crossing any fixed point in its path per second. The frequency of the radiation can be obtained by dividing the velocity of light by the wavelength. Thus the longer the wavelength, the shorter the frequency and vice versa.

A star or a galaxy gives off electromagnetic radiation in all wavelengths. Radiation of different waves arises from different mechanisms in the star or galaxy. For example, because of nuclear burning in the star (this will be explained in detail later) a great deal of heat and light is generated in the star. It gives off this radiation and gradually cools down. Then there is the radiation in the radio band from the motion of electrically-charged particles such as electrons and protons of which all matter is made. The radiation carries off energy to the surrounding space from these charged particles, which thereby lose energy. Actually, in the ultimate analysis all radiation is due to the motion of charged particles. When we heat a piece of iron we essentially increase the random motion of the electrons in it, which produces heat or infra-red radiation. As this random motion increases we get radiation of higher frequency, namely light, that is, the iron becomes 'red hot' and so on.

The radiation from any source has different amounts of energy in different wavelengths. That is, the intensity of the radiation may be different at different wavelengths. Thus one can draw a graph of the intensity versus the frequency which may look something like the graph of Fig. 3.5. This graph

Table 3.1. *Wavelength and nature of radiation*

Nature of radiation	Wavelength (cm)
Radio (up to VHF)	greater than 10
Microwave	0.01–10
Infra-red (heat)	0.0001–0.01
Visible light	0.00002–0.0001
Ultra-violet	10^{-7}–0.00002
X-ray	10^{-9}–10^{-7}
Gamma ray	less than 10^{-9}

shows that the highest intensity occurs at the wavelength λ_0. Such a graph is referred to as the 'spectrum' of the radiation. The word 'spectrum' also refers to the resolution of light from a source into its constituent colours with the help of a spectrometer, a simple form of which is the prism. Different radiating objects such as stars and galaxies and pieces of iron have different characteristic spectra. Some of the radiation from a star or a galaxy is usually absorbed by colder gas clouds in the outer regions of the star or galaxy. This absorption occurs at certain definite wavelengths depending on the nature of the matter that is absorbing the radiation. Thus, for example,

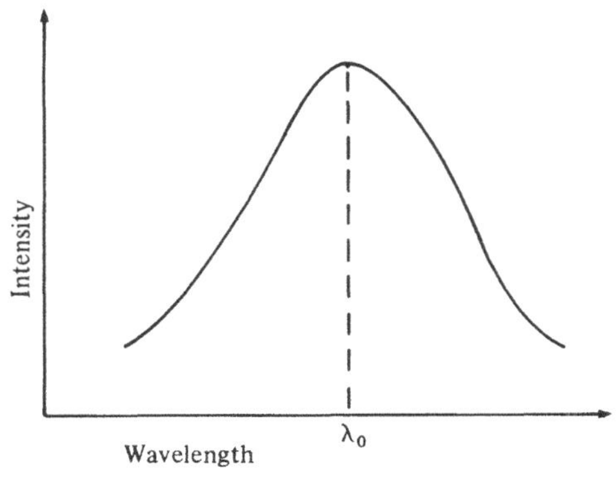

Fig. 3.5. A graph of the intensity versus wavelength for the radiation from a source. Maximum intensity occurs at the wavelength λ_0.

calcium atoms absorb radiation at a particular wavelength, iron atoms at another wavelength and so on. The wavelengths at which different materials absorb radiation are well known from laboratory studies. The absorption of radiation at certain wavelengths in the outer regions results in dark lines at these wavelengths in the spectrum of light received from the star or galaxy. Hubble studied light from distant galaxies and found that he could recognize in them dark lines as caused by known forms of matter only if he assumed that these dark lines had been systematically shifted to higher wavelengths. By carefully studying the spectra of a large number of galaxies, he came to the conclusion that these shifts (called 'red shifts', because for visible light the shift is towards the red end of the spectrum) are due to the recessional velocities of the galaxies.

The appearance of dark lines in the spectrum of light from astronomical bodies was actually known from much earlier times. The German physicist Joseph von Frauenhofer (1787–1826) found these dark lines in the spectrum of the light from the sun. Dark lines in the spectrum of radiating objects had also been observed by the English chemist William Hyde Wollaston (1766–1828) in 1802. In 1868 the English astronomer William Huggins (1824–1910) showed that the dark lines in the spectra of some of the brighter stars are shifted systematically towards the red or the blue from their normal position in the spectrum of the Sun. He correctly interpreted this as a Doppler effect, due to the motion of the stars towards us or away from us. Thus the wavelength of the dark lines in the spectrum of the star Capella is longer than the corresponding wavelength in the spectrum of the Sun by 0.01% towards the red. This indicates that Capella is moving away from us at 0.01% of the velocity of light, that is, 30 km/s. In the following decades the Doppler effect was used to find the velocities of various astronomical bodies such as double stars, the rings of Saturn, etc.

How did Hubble know that the galaxies with higher red shifts (and higher recessional velocities) are the more distant galaxies? This is because he found that on the average the fainter the galaxy, the higher the red shift. Now in general the fainter galaxies are the more distant ones. One has to be careful here because the faintness of a galaxy may be caused not only by its distance, but also it may be giving off less radiation than others, that is, it may be intrinsically a less luminous galaxy. For this Hubble had to make a study of the various types of galaxies and choose a certain class of galaxy which, independent of the distance from us, gives off roughly the same total amount of radiation, that is, has the same absolute luminosity. These galaxies are known as 'standard candles' because one can infer their distance from their faintness, that is from their apparent luminosity. The problem of finding 'standard candles' is a very difficult one which has not yet been completely solved. Thus what Hubble really found was a relation between the red shift and apparent luminosity of distant galaxies. From the above reasoning one can interpret this as a relation between the velocity of recession and distance of galaxies. This is really an extrapolation from observation.

The red shift can be caused by other processes than by the velocity of recession of the source. For example, it is known that if light is emitted by a source in a strong gravitational field and received by an observer in a weak gravitational field then the observer will notice a red shift of the light. However, it seems unlikely that the red shift of distant galaxies is gravitational in origin; for one thing these red shifts are rather large for them to be gravitational and, secondly, it is difficult to understand the systematic increase of the red shift with faintness on the basis of a gravitational origin. Thus the present concensus of opinion among experts is that the red shift is due to velocity of recession, but an alternative explanation of at least a part of these red shifts on the basis of either gravitation or some hitherto unknown physical process cannot be completely ruled out.

Hubble's Law implies arbitrarily large velocities of the galaxies as the distance increases indefinitely. Does not this violate the Special Theory of Relativity? Astronomers usually denote the size of the red shift by the letter z. It is the fractional shift of the wavelength, that is, the difference between the received wavelength and the original wavelength, this difference being divided by the original wavelength. For velocities low compared to the speed of light, the velocity of the galaxy is just its red shift z multiplied by the velocity of light c, that is, its speed is cz. Thus if the galaxy has a red shift of 0.15, then its velocity is 15% of the speed of light and so on. For velocities greater than about one-third of the velocity of light, the simple relation between speed and red shift does not hold. Although it is possible to observe very high red shift, it is not possible to *observe* velocities higher than the velocity of light. In fact as the red shift approaches infinity, the corresponding velocity approaches the speed of light. This relation is illustrated by the graph in Fig. 3.6. The distance at which the red shift from a galaxy becomes infinite is usually referred to as the 'horizon'. Thus galaxies beyond the horizon are not observable. Is it then not the case that galaxies which are beyond the horizon have a velocity with respect to us which is faster than light? In some sense they do, but this does not violate the Special Theory of Relativity for several reasons. For one thing the Special Theory

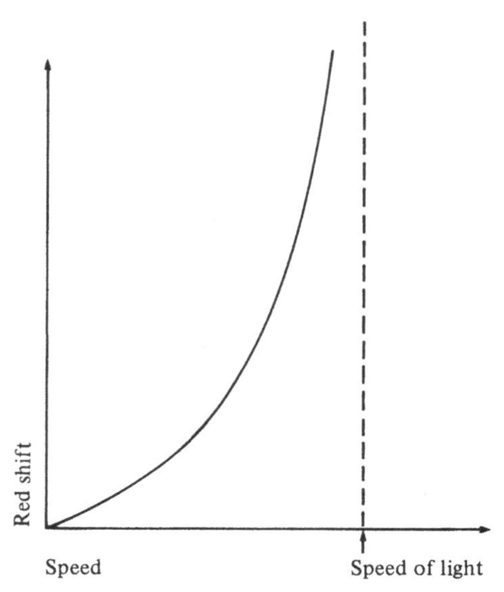

Fig. 3.6. This graph shows the relation between the red shift (z) and the speed of recession. As z tends to infinity, the speed of recession tends to the speed of light.

is valid only in the absence of gravitation, whereas in the universe there is gravitation present everywhere. This gravitational field radically alters the nature of space and time according to the General Theory of Relativity of Albert Einstein (1879–1955). It is not as if a material particle is going past an observer at a velocity greater than light, but it is space itself which is in some sense expanding faster than the speed of light. As mentioned earlier, galaxies beyond the horizon are not directly observable although we can infer their existence. Gravitation introduces 'curvature' into space, which alters the concept of velocity between two observers who are very far from each other in a 'curved' space. The best way to define the velocity of a distant galaxy is through its red shift. So all one can say is that the red shift becomes infinite at the horizon, beyond which galaxies are at present unobservable. The difficulties considered in this paragraph can be elucidated by a precise mathematical formulation in terms of the General Theory of Relativity, but it is not necessary to understand the subtleties of this particular problem in order to follow the main line of argument in this book.

Models of the universe

It is appropriate at this stage to consider the concept of a 'model' of the universe. A model of the universe is a hypothetical universe which incorporates all the known properties of the observable universe and some plausible assumptions derived from observations. A model is often used in science to test hypotheses about certain objects or processes. The model can be used to make predictions which one can then attempt to verify by observation or experiment. The model also helps to make a conceptual picture of the object or physical process which may lead to a better understanding of it. There is no sharp dividing line between 'model' and 'theory'. The term 'model' is used when one makes a construction, either real or conceptual. For the *universe as a whole* it is necessary to have a model in mind since, unlike objects or processes in the laboratory, we cannot observe it in its totality. One of the first persons to consider a model or theory of the universe as a whole was Isaac Newton (1642–1727). It was natural for him to apply the laws of dynamics and the law of universal gravitation which he had formulated to the universe as a whole. These laws had very successfully explained the motion of the planets. In a letter to the classical scholar Richard Bentley (1662–1742), Newton writes 'But if the matter were evenly disposed throughout an infinite space . . . some of it would convene into one mass and some into another, so as to make an infinite number of great masses, scattered great distances from one another throughout all that infinite space. And thus might the sun and fixed stars be formed, supposing the matter were of a lucid nature.' He was also aware of the difficulty of having a universe with the matter distributed evenly in a *finite* region, for he realized that then the matter would all tend to fall towards the centre, 'and there compose one great spherical mass'.

It has turned out that Newtonian dynamics and gravitation are inadequate to provide a theoretical framework for considering the universe as a whole. But this was not the whole reason why Newton and his successors were unable to deal adequately with cosmology. They made an assumption about the universe which seemed obvious at the time but which has turned out to be untrue, namely, they assumed that the universe as a whole is a *static* system with no large-scale changes taking place. As we have already seen, and will see more clearly later, the universe is a *dynamic* system. It has turned out that Einstein's General Theory of Relativity provides a suitable mathematical and conceptual framework for describing the universe as a whole. But it was shown in 1934 by the English astrophysicist Edward Arthur Milne (1896–1950) in collaboration with W.H. McCrea, that many of the results of relativistic cosmology (i.e. cosmology based on the General Theory of Relativity) can be derived using Newtonian ideas provided the assumption is made that the universe is in a dynamic state. Thus progress in cosmology need not have been delayed till the advent of the General Theory of Relativity if the assumption that the universe is static had been abandoned. There arose a good reason for abandoning this assumption in 1826 with the so-called Olbers' paradox, which unfortunately was largely ignored.

The German astronomer and physician Heinrich Wilhelm Matthaus Olbers (1758–1840) took as a basis of his paradox the remarkably simple observation that the sky is dark at night. He assumed that the universe was infinite and static, that is, on the average the relative velocity between two stars vanished. He also assumed that the average density of stars and the average absolute luminosity of stars were constant throughout the universe, provided these averages were taken over sufficiently large regions. He also implicitly assumed that space was Euclidean, that is, Euclid's geometry was satisfied everywhere in space. He then showed that these plausible assumptions led to a contradiction, as follows. Consider a large spherical shell whose centre is some arbitrary point O of space. Let this shell have inner radius r and thickness h, and suppose that r is much greater than h. Then the surface area of the inner surface of the shell is $4\pi r^2$ (here π stands for a number approximately equal to 22/7) and the volume of the shell can be taken approximately to be $4\pi r^2 h$. Now let L be the amount of light emitted per unit volume, that is, L is obtained by multiplying the number density of stars by the absolute luminosity of each star. Thus the light emitted by the spherical shell is given by $4\pi r^2 hL$. Now the intensity of light from a source decreases inversely as the square of the distance, that is, if the distance is doubled, the intensity becomes a quarter, if the distance is trebled, the intensity becomes one-ninth and so on. Using this fact, it can be shown that the intensity of light at the centre O due to the stars in the spherical shell is given by hL so that it is independent of the radius of the shell. If we surround our shell by a series of shells of equal thickness concentric with the first, the outer boundary of each shell being the inner boundary of the next, then each shell will make the same contribution to the radiation density of the centre O. Since we can add on shells indefinitely, it follows that the radiation density is infinite at the centre. In this analysis we have ignored the fact that light from a star may be absorbed by another on its way to the centre O. When this fact is taken into account it can be shown that the radiation density at O should be the same as at the surface of an average star. Thus from a few simple assumptions one can derive the remarkable result that the sky should everywhere be as hot as the surface of a star. An assumption implicit in this analysis is that the universe should have existed for an infinite time in the past, for the radiation to have had the time to travel large distances and for thermal equilibrium to be established. In fact the same conclusion as above can be reached by considering the fact that a static system of infinite age must have reached thermodynamic equilibrium, with each star absorbing as much radiation as it emits.

Since the conclusion reached above is contrary to observation, one of the assumptions in the analysis must be wrong. In fact it has turned out that several of the assumptions are wrong. The universe is not static and it has not existed for infinite time, at least not in the sense that Olbers assumed. In an expanding universe light from distant galaxies gets red shifted and so loses energy, so the intensity of light decreases faster than inversely with the square of the distance, contrary to the above analysis. If Olbers' paradox had been taken seriously at the time it was proposed, people would have been encouraged to question the assumptions under which it was derived, and at the very least the expansion of the universe would not have come as a great surprise as it did after Hubble's discovery.

I shall now describe in some detail the model of the universe that is currently in favour, that is the standard model. It is actually a collection of several related models in the sense that there is as yet no unique model that fits all the observations. It will seem as if I am describing the actual universe but in reality I shall be concerned with a model, in the sense that many of the properties I shall describe are inferred from various observations and arguments based on observations. Some of these arguments and observations have already been cited.

The universe, as we have seen, appears to be homogeneous (that is, it has a uniform distribution of galaxies) and isotropic as far as we can detect. These properties lead us to make an assumption about the model universe, called the Cosmological Principle. According to this Principle the universe is homogeneous everywhere and isotropic about every point in it. Again, the homogeneity and isotropy are to be considered in an average sense. This assumption is very important, and it is remarkable that the universe seems to obey this Principle. The Cosmological Principle makes it possible to study the universe as a single evolving entity. This Principle asserts that the universe is not simply a random collection of irregularly distributed galaxies, but it is a single entity, all parts of which are in some sense in unison with all other parts.

The Cosmological Principle simplifies considerably the study of the large-scale structure of the universe. It implies, amongst other things, that the distance between any two typical galaxies has a universal factor, the same for any pair of galaxies in the following sense. Consider any two typical galaxies A and B which are partaking of the general motion of expansion of the universe. The distance between the two galaxies can be obtained by multiplying a number f_{AB} (this number depends on A and B) by another number R, that is, the distance is given by $f_{AB}R$. Here the number f_{AB} does not change with time, but the number R changes with time, in the sense that it has different values at different epochs in the history of the universe. In mathematics this is expressed by saying that R is a function of time but f_{AB} is independent of time. It is one of the consequences of the Cosmological Principle that the distance between *any* two typical galaxies has the same form. For example, the distance between galaxies C and D which are partaking of the expansion has the form $f_{CD}R$, where f_{CD} depends only on the galaxies C and D and does not change with time. One of the consequences of this result is that if the distance between the galaxies A and B doubles in a certain period of time then the distance between galaxies C and D also doubles in this period of time. In fact the distance between *any* two typical galaxies doubles in this period of time. The large-scale structure and behaviour of the universe can be described by the single quantity R, which, as mentioned earlier, changes with time. One of the major current problems of cosmology is to determine how R changes with time. The quantity R is called the scale factor or the radius of the universe. The latter term is somewhat misleading, because, as we shall see, the universe may be infinitely large in its spatial extent in which case it will not have a finite radius. However, it may also turn out that the universe is finite in spatial extent, in which case R is related to the maximum distance between two points in the universe. Because of this uncertainty I prefer to call R the *scale factor*.

Let us try to understand in some detail what the recession of the galaxies, as discovered by Hubble, means for the universe as a whole. In this it is helpful to consider the analogy of a spherical balloon which is expanding and which is uniformly covered on its surface with dots (see Fig. 3.7). In this analogy we should concentrate to begin with on the surface of the balloon and ignore the surrounding space in which the balloon exists, or in which it is embedded. We may suppose that the surface of the balloon is inhabited by two-dimensional creatures who cannot leave the surface; their whole universe

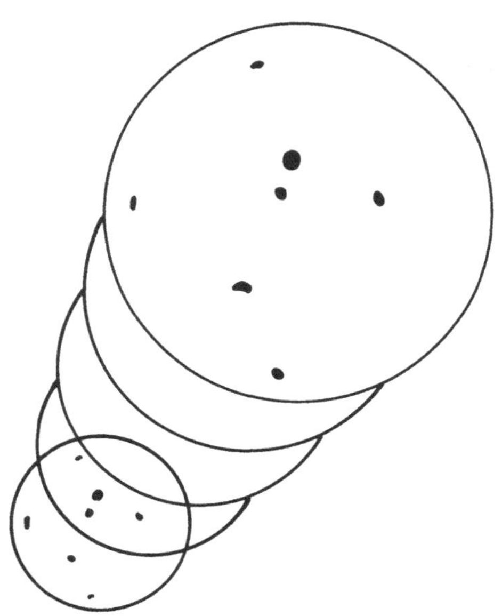

Fig. 3.7. The expanding universe is analogous to an expanding balloon with dots on its surface. At any point on the balloon's surface the dots appear to be moving away at speeds proportional to their mutual distance.

is the surface of the balloon. The dots can be considered to correspond to galaxies. As the balloon expands, all dots move away from each other. From any given dot all dots appear to move away from it with speeds which at any given time are proportional to the distance (along the surface) from the given dot. This property, which can be easily established, corresponds to Hubble's Law. In other words, suppose an observer is placed in one of the dots. Then the velocity away from him of any other dot is found by multiplying the distance of the dot along the surface by a certain number, which is the same number no matter which dot he is considering. This number corresponds to the Hubble constant. The appearance from all dots will be the same; in a sense every dot can be considered to be the centre of expansion. Thus it is clear that although all dots move away from each other, there is no 'central' dot with a privileged position. In a similar manner, the fact that all galaxies are moving away from our galaxy does not imply that our galaxy has a privileged position, rather we would observe the same recession no matter which galaxy we were situated in.

The analogy between the universe and the surface of a uniformly-dotted expanding balloon can be taken further. Suppose we have two dots A and B on the surface. Join the points A and B by straight lines to the centre of the spherical balloon, which we denote by O (see Fig. 3.8). Let the angle AOB be called θ_{AB} and let it be measured in radians (an angle of 90°, that is, a right angle, is $\frac{1}{2}\pi$ radians where, as mentioned earlier, π stands for a number which is approximately 22/7). Then it is a simple geometric fact that the distance between A and B along the surface, that is, along a great circle (a great circle is the circle in which a plane through the centre of the sphere intersects the surface of the sphere) is found by multiplying θ_{AB} by the radius of the balloon. Let us call R' the radius of the balloon. Since the radius is changing, R' will be different at different times. Thus the distance between the dots A and B along the surface is given by $\theta_{AB}R'$. Now as the balloon expands the angle θ_{AB} remains the same, as it depends only on the fixed dots A and B (since the surface expands uniformly, the angle subtended at the centre by any two given dots remains the

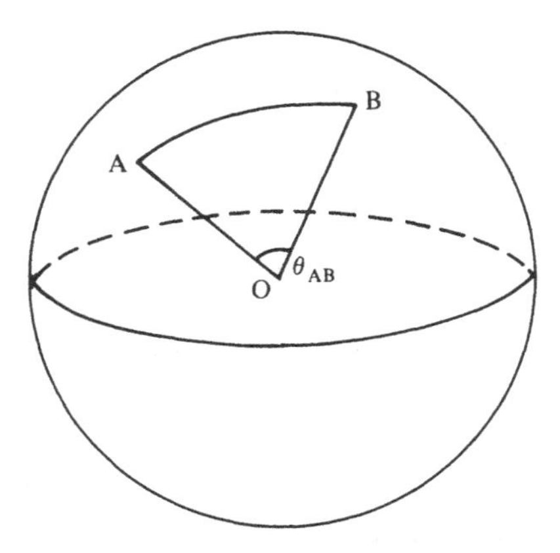

Fig. 3.8. Diagram to illustrate the distance between two points on a sphere and the angle which they subtend at the centre.

same). Also the distance along the surface between any pair of dots has the same form, namely $\theta_{CD}R'$, where θ_{CD} is the angle subtended at the centre by dots C and D, and remains fixed in time. The reader will have noticed the similarity between the form of the distance between two dots and the form of the distance between a pair of galaxies mentioned earlier. This discussion can be extended to show that the speed of recession of dots is proportional to the distance, just like Hubble's Law for galaxies. In the case of galaxies Hubble's Law is only approximately true but for dots on a balloon the corresponding relation is strictly true. From the form of the distance between two dots it also follows that if the distance between typical dots A and B doubles in a certain period of time, the distance between *any* two typical dots doubles in the same period of time.

Of course in some respects the example of the uniformly-dotted expanding balloon differs from the universe. For example, the dots on the balloon are on a two-dimensional surface at any given time, whereas galaxies are spread over a three-dimensional space. Also, unlike the surface of the balloon the universe may be infinite in spatial extent. However, it may also be finite. It is not known at present whether the universe is infinite or finite in spatial extent. If the universe is finite, it is not as if galaxies exist up to a certain boundary in space beyond which is empty space. Rather, in a finite universe it is still true that galaxies are spread uniformly throughout *all* space but the space 'closes on itself' so that if we could draw a 'straight' line, the 'straight' line would eventually return to the same point from the opposite direction. The total distance along this straight line may be called the circumference of the universe and it is finite. We cannot build a model of a finite universe in the laboratory by joining rods to represent distances, because in the laboratory the rods necessarily satisfy Euclidean geometry whereas in the scale of the universe (in the finite case) the geometry is non-Euclidean. An example of the non-Euclidean nature of space will be given below. I have put the word 'straight' above within quotes because in a non-Euclidean space the nearest thing to a straight line is what is called a geodesic and it is by following a geodesic in a finite universe that we come back to the same point from the opposite direction. (For example, on the surface of a sphere the geodesics are great circles; we cannot draw a straight line on such a surface.) Thus a finite universe has no boundary (nor, of course, does the infinite universe). In this sense it does resemble the surface of a sphere which in two dimensions is finite but unbounded, that is, it has no boundary. In a finite universe the amount of matter is finite whereas if the universe is infinite, which is also theoretically possible, we have an infinite amount of matter filling an infinite space. Neither possibility is easy to comprehend even for experts, but these models of the universe can be represented mathematically within the framework of the General Theory of Relativity and this mathematical framework helps to make a conceptual picture of these possibilities. Of course the ultimate test of whether these models are correct is that physical consequences derived from them must agree with observations. At present both the finite and infinite models of the universe are tenable, because observations cannot decide between them.

From the rate at which galaxies are receding away from each other, it can be deduced that *all* galaxies must have been very close to each other *at the same time* in the past. Consider the balloon analogy again; from the rate of expansion of the balloon one can trace the expansion backwards and deduce that the balloon must have started at some time with *zero radius R′* and at this initial time all the dots were on top of each other *at the same time.* Similarly, from the expansion of the galaxies we assume that the galaxies must have been 'on top of each other' at some initial moment when the scale factor R was zero. It is believed that at the initial moment (sometime between 10 and 20 billion years ago) there was a universal explosion in which matter was thrown asunder violently. This explosion is referred to as the 'big bang'. This was a most unusual explosion in the sense that it was an explosion *at every point of the universe.* Of course, as mentioned earlier, this can mean every point of an infinite universe or every point of a finite universe. If the universe is finite, it would have started from zero volume. But it *must not be supposed* that in this case the matter was concentrated in a small volume in an otherwise empty space and that it exploded, spreading into the surrounding empty space. In fact there is no 'outside' or 'inside' in a finite universe; the whole of space is finite in this case. Again, this is not easy to comprehend, but the mathematical models help to form a conceptual picture and imply the properties we have been describing. An infinite universe remains infinite all the time down to the initial moment; as in the case of the finite universe, the matter becomes more and more dense and hot as one traces the history of the universe to the initial moment, which is usually referred to as a 'space–time singularity' or simply a 'singularity'. Very little is known about the precise nature of the initial singularity or the big bang.

The universe is expanding now, that is, the galaxies are receding away from each other because of the initial explosion. There is no force propelling the galaxies apart, but their motion is simply the remnant of the initial impetus. In fact, the recession is slowing down because of the gravitational attraction of different parts of the universe to each other.

Is there any direct evidence for the big bang? There is an important piece of evidence apart from the recession of the galaxies, that the contents of the universe in the past must have been in a highly compressed form. This is the so-called cosmic background radiation, whose existence can be explained as follows. As we trace the history of the universe to the past, the matter in the universe becomes more dense. At some stage in the past, galaxies could not have had a separate existence, but must have been merged together to form one great continuous mass. Now as matter is compressed, its temperature rises, so the matter in the universe must have been at a high temperature in the past. In fact there is reason to believe, as we shall see later, that there must have been a great deal of electromagnetic radiation (radio waves, infra-red waves, light, etc.) in addition to matter, and this radiation at some stage must have been in equilibrium with matter. When radiation is in equilibrium with matter, that is, when matter absorbs and emits equal amounts of radiation in every wavelength, then the spectrum of radiation (the graph of its intensity or energy density versus wavelength) has a particular form, which depends only on the temperature of the matter and not on the nature of the matter with which the radiation is in equilibrium. This spectrum looks something like that of Fig. 3.5 for ordinary temperatures and is referred to as 'black-body radiation' because it corresponds to the spectrum of the radiation emitted by a perfectly black body. The problem of finding the spectrum of black-body radiation is historically very important because it gave rise to the quantum theory and reveals the fact that radiation comes in discrete chunks of energy, later called photons (more about quantum theory and photons later). The problem was solved by the German physicist Max Karl Ernst Ludwig Planck (1858–1947) at the turn of the century. Planck's formula can be stated quantitatively as follows: in a box filled with black-body radiation, the energy or intensity of the radiation in any range of wavelengths rises very steeply with increasing wavelength, reaches a maximum and then falls off steeply again. According to the current view of the early universe there should be a remnant of the radiation that was in equilibrium with matter at high temperature at early times, but this radiation should have a much lower temperature now, because as the universe expands the radiation cools. It can be shown that the spectrum of the radiation should continue to have the black-body form. Thus at the present time there should pervade the whole universe electromagnetic radiation with a black-body spectrum and with a certain low temperature; this radiation should be homogeneous and isotropic like the universe itself. Such an isotropic radiation was indeed first discovered by A.A. Penzias and R.W. Wilson in 1965. This radiation was found to have a temperature of approximately 3 K (K stands for Kelvin, a scale which measures absolute temperature, that is, the temperature upwards from absolute zero, which is approximately -273° centigrade). Penzias and Wilson found the radiation at a wavelength of 7.35 cm; their discovery has since been confirmed by many other observers working at different wavelengths. So far the spectrum of this isotropic background radiation (it is called 'background' because it comes from nowhere in particular but pervades the whole universe) is found to be of the form expected for black-body radiation, although more results are needed to establish this very firmly. The spectrum peaks (the wavelength λ_0 in Fig. 3.5) at slightly below 0.1 cm. It is difficult to get measurements for wavelengths less than 0.1 cm (in the infra-red range) as the atmosphere is opaque to such radiation so that observations have to be done above the atmosphere in satellites. There are indications that the graph curves over as in Fig. 3.5, as it should. There have been some attempts to explain the cosmic background radiation in terms of some other sources than the remnant of an early phase of the universe. For example, this radiation could have arisen due to the radiation from galaxies all over the universe. Such explanations, however, have not been very successful.

A slight systematic anisotropy is to be expected in the cosmic background radiation. This is because as the earth travels through the radiation, it should appear to be very slightly warmer in the direction in which the earth is travelling and very slightly cooler in the opposite direction. Such a systematic variation in the isotropy was indeed discovered in 1977 by R.A. Muller and his collaborators. They found that the temperature was slightly higher (by about one-three hundredth of a degree) in the direction of the constellation Leo and cooler by the same amount in the opposite direction. This indicates the absolute motion of the Earth through the universe; it is moving towards the constellation Leo at a speed of about 400 km/s. Actually the Earth goes round the Sun and the Sun goes round the centre of the galaxy. When all this motion is taken into account one comes to the conclusion that our galaxy is moving through space at about 600 km/s. This speed is rather large but not inexplicable. It could arise for example because of the rotation of our Galaxy around the local cluster of galaxies.

The observation of the cosmic background radiation is one of the two most important observations in cosmology, the other being Hubble's discovery of the recession of galaxies. The existence of the cosmic background radiation is a strong indication that the universe has gone through a hot and dense early stage in which matter and radiation were in equilibrium. This evidence therefore supports the 'big bang' origin of the universe. However, it would be premature to say that the problem of the origin of the universe has been solved. Much remains to be understood about the precise nature of the big bang. In particular, was there a stage of the universe before the big bang, or did time have a beginning at the moment of the big bang? These are among the most difficult questions in science to which no satisfactory answers exist at present.

4

Elementary particles – a preliminary look

In this chapter we shall digress and take a first look at some of the elementary particles and their properties, knowledge of which will be useful in several places in the following chapters. We shall take a more detailed look at this subject in Chapter 14, when we consider the important question of the stability of the proton. Consider first the particle associated with light or electromagnetic waves. An alternative description of radiation exists in terms of particles called *photons*. It was realized at the turn of the century by Planck and later by others that radiation consists of discrete chunks of energy which are called photons. This is one of the consequences of the quantum theory, about which we will learn more later. Photons have most of the attributes of particles, and they can be considered as such. An ordinary light wave consists of billions of photons travelling all together but if we were to measure the energy of the wave very precisely we would find that it is a multiple of a definite quantity, which can be considered as the energy of a single photon. The energy of a photon is usually quite small so for most practical purposes the energy of an electromagnetic wave can have any value. However, the interaction of light or electromagnetic wave with an atom or atomic nucleus takes place one photon at a time. It is important to consider the photon picture when considering these interactions. Considered as particles, photons have zero mass and zero electrical charge but they have energy depending upon the frequency of electromagnetic wave with which they are associated. According to the Special Theory of Relativity all particles of zero mass travel at the speed of light, unlike massive particles.

The question as to what constitutes an elementary particle is itself a fundamental problem in theoretical physics and we need not concern ourselves with it. I will just mention some relevant particles which can be considered as elementary for our purpose. The first of these, the *electron*, was discovered in 1897 by the English physicist Joseph John Thomson (1856–1940). The electron has one negative unit of electric charge. Electricity consists of the flow of electrons. The electron mass is about 9.11×10^{-28} g. The *proton* was discovered around 1920 by the New Zealand physicist Ernest Rutherford (1871–1937). The proton electric charge is equal and opposite to that of the electron, that is, it has one positive unit of electric charge. The proton is 1836 times as massive as the electron. The *neutron* was discovered by the English physicist James Chadwick (1891–1974) around 1932. The neutron is 1838 times more massive than the electron and carries no electric charge. An atom consists of a nucleus made up of neutrons and protons (a hydrogen nucleus is a single proton) confined to a size of about 10^{-13} cm. Because of their opposite charge, electrons and protons attract each other. Around the nucleus are electrons, the same number as the number of protons in the nucleus so that an atom has no net charge, that is, it is neutral. Atomic dimensions are about 10^{-8} cm. Thus all ordinary matter consists of electrons, protons and neutrons. Following the work of the English physicist P.A.M. Dirac in the late 1920s it became known that for every particle there exists an *anti-particle* with the same mass but opposite charge and opposite values of some other attributes which we need not consider. When a particle and an antiparticle come together, they annihilate each other and give off a burst of radiation. The energy E contained in this radiation is given by Einstein's celebrated formula $E=mc^2$, where m is the combined mass of the pair and c is the velocity of light. The antiparticle of the electron is the *positron* discovered in 1932 by the American physicist C.D. Anderson. The antiparticles of the proton and neutron are, respectively, the *antiproton* and *antineutron*. The photon can also be considered as an elementary particle; it is its own antiparticle. Most elementary particles possess an intrinsic property called *spin*. To picture this property, one can imagine the particle to be like a little sphere which is spinning or rotating about an axis. This picture should not be taken too literally. The spin results in the particle having *angular momentum*, which bears the same relation to rotatory motion as ordinary momentum does to motion from one point to another point. According to the quantum theory, angular momentum comes in discrete amounts, measured by a fundamental unit given by Planck's constant $\hbar$ (this is read as h-slash and is equal to Planck's original constant h divided by 2π). In terms of this unit, angular momentum or spin can take integral or half-odd-integral values (that is $0\hbar$, $\frac{1}{2}\hbar$, $1\hbar$, $1\frac{1}{2}\hbar$, $2\hbar$ etc.). We do not notice this discreteness in everyday life because the unit $\hbar$ of angular momentum is an exceedingly small amount. For example, a child's toy top when normally spinning probably has about $10^{30}\hbar$ units of angular momentum. The picture of a rotating sphere for a particle is a little misleading in the sense that the angular momentum from such a system is referred to as 'orbital' angular momentum and these can take only *integral* values according to quantum mechanics. The electron, for example, has spin $\frac{1}{2}\hbar$ which is not integral. Thus it is a somewhat peculiar kind of intrinsic spin that elementary particles have which cannot be pictured easily in everyday terms. Protons and neutrons also have spin $\frac{1}{2}\hbar$. The photon has unit spin ($1\hbar$) but it should not be regarded as arising out of 'orbital' angular momentum even though it is integral.

We now come to the elusive particle called the *neutrino* (meaning 'little neutral one'). Like the photon it has zero mass (at least this was the assumption until recently), and no electrical charge, but unlike the photon it has spin $\frac{1}{2}\hbar$. It has very little interaction with ordinary matter. In fact it can pass right through several light years of lead without being stopped. The existence of the neutrino was proposed in 1931 by the Austrian physicist Wolfgang Pauli (1900–1958) and the neutrino was finally detected experimentally by F. Reines and C. Cowan in 1956. It turns out that a free neutron (that is, a neutron which is not in a nucleus) is unstable in the sense that in a few minutes it decays into a proton, an electron and an antineutrino (this is the antiparticle of the neutrino which is like the mirror image of the neutrino – see Fig. 4.1). This process is called beta decay, and is historically very important because it is one of the first-known examples of a new kind of interaction known as the weak interaction. We shall come back to this later. In fact there is more than one type of neutrino. The electron has associated with it one kind of neutrino called the *electron-neutrino.* There exists another type of particle in nature called the *muon* which has the same properties as the electron but is about 207 times heavier than the electron. The muon is negatively charged and its antiparticle (corresponding to the positron) is positively charged. The muon has associated with it *its* neutrino called the *muon-neutrino.* Since in beta decay an electron is involved, the antineutrino that is produced in this decay is an electron-antineutrino. The electron, the muon, the neutrinos (and their antiparticles) belong to a family of particles called *leptons.* The significance of this property is that it is possible to assign a *lepton* number to each of these particles in such a way that the total lepton number before a process takes place is the same as the total lepton number after the process. Recently a new kind of lepton has been discovered called the τ-lepton ('τ' is the Greek letter tau). Presumably the τ-lepton also has neutrinos associated with it which we shall call the τ-neutrinos.

The masses of elementary particles are not usually given in grams as 1 g is a large mass for elementary particles. Instead they are given in terms of the energy that would be produced if that amount of mass were totally converted into energy using

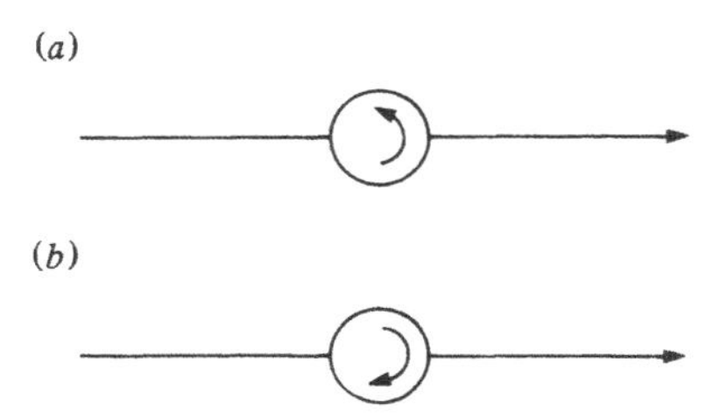

Fig. 4.1. The neutrino spins like a left-handed screw (*a*), whereas the antineutrino is like the mirror image (*b*).

Einstein's formula $E = mc^2$. This energy is expressed in terms of the *electron volt*, written as ev. One electron volt is given by 1.602×10^{-12} ergs. The kinetic energy (energy due to motion) of a mass of 1 g travelling at a speed of 1 cm/s is half an erg. Thus the electron volt is a minute amount of energy by everyday standards. The mass of an electron is 0.51 million electron volts (Mev). The masses of the proton and neutron are, respectively, 938.3 Mev and 939.6 Mev.

5

Is the universe open or closed?

Will the expansion of the universe continue forever or will it at some future time stop expanding and start to contract? This is one of the most important unsolved problems in cosmology. In terms of the scale factor R of the universe, this question amounts to asking whether R will increase with time for all future times or whether it will reach a maximum, then decrease and finally reduce to zero. These two possibilities are represented in Fig. 5.1 by the curves marked 'open' and 'closed', respectively. The open universe expands forever, that is, typical intergalactic distances keep on increasing forever, while in the closed universe typical distances between galaxies reach their maximum value, then start decreasing, to reach zero in a finite time. If one restricts oneself to the simple models of the universe (these are called Friedmann models after the Russian mathematician Alexander Alexandrovitch Friedmann (1888–1925) who found these models as solutions of Einstein's equations), then the open universe is infinite in spatial extent and the closed universe is finite.

For the Friedmann models the geometry of space in the infinite universe is different from that in the finite universe in the following sense. In the ordinary space that we are used to (this is called Euclidean space because Euclid's geometry is valid in this space) if we have a sphere of radius r, its surface area is $4\pi r^2$ and its volume is $\frac{4}{3}\pi r^3$. In the space of the closed (finite) universe a sphere of radius r has a surface area less than $4\pi r^2$ and volume less than $\frac{4}{3}\pi r^3$. This space is called 'spherical space'. In the infinite (open) universe there are two possibilities. In the first case the geometry is the same as in Euclidean space.

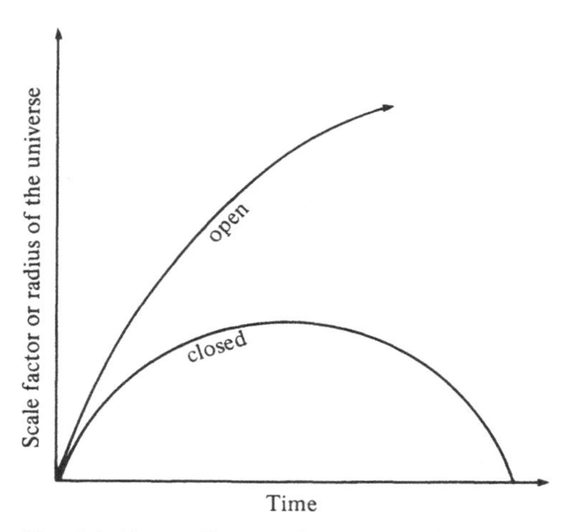

Fig. 5.1. Depending on the density of matter in the universe, gravity may eventually halt the present expansion of the universe and cause it to collapse, or the universe may expand forever. The former corresponds to the closed model, and the latter to the open model.

In the second possibility a sphere of radius r has surface area greater than $4\pi r^2$ and volume greater then $\frac{4}{3}\pi r^3$. This is called hyperbolic space. We can find two-dimensional analogues of these spaces by considering a circle in a plane, on the surface of a sphere and on the surface of a hyperboloid of one sheet (see Fig. 5.2). The circumference and area of a circle of radius r in a plane are, respectively, $2\pi r$ and πr^2. On the surface of a sphere a circle of radius r (the radius has to be measured from the centre of the circle on the sphere along a great circle to the circumference) has circumference less than $2\pi r$ and area less than πr^2. On the hyperboloid of one sheet the 'circle' of 'radius' r has circumference greater than $2\pi r$ and area greater than πr^2. For any surface the circle is to be interpreted as a curve whose 'distance' along the surface from a fixed point (the centre) remains the same. Here the 'distance' is along the shortest curve lying *on the surface*, which, as mentioned earlier, is called a geodesic. These analogies with two-dimensional surfaces explain why the geometry in the closed universe is called spherical and the two types of geometries in the open universe

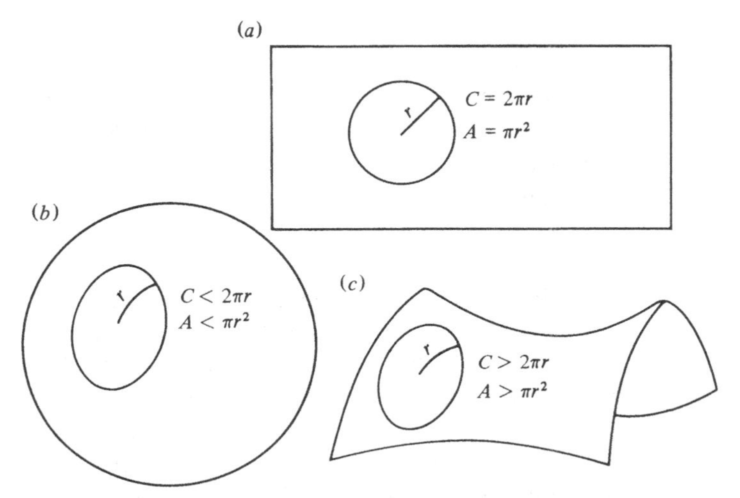

Fig. 5.2. The circumference C and area A of a 'circle' (*a*) in a plane, (*b*) on the surface of a sphere and (*c*) on the surface of a hyperboloid of one sheet.

are called Euclidean (flat) and hyperbolic, respectively. The surface of a hyperboloid of one sheet is not a good analogy of the hyperbolic space in one sense, in that the former has a centre whereas the latter does not.

By making measurements of the surface area of a sphere of radius r, in principle it is possible to determine the geometry of space, and hence to determine whether our universe is open or closed (assuming that it is a Friedmann universe). However, in practice it is completely beyond present technology to make such measurements. The geometry of space manifests itself in large-scale properties of space and is not of any consequence when considering physical processes occupying relatively small portions of the universe. By 'small' we mean small compared with distances in which the universe appears homogeneous. Even if the geometry of the universe were spherical or hyperboloid, in scales such as that of our Galaxy the geometry would be almost Euclidean and it would be exceedingly difficult to measure the minute departure from this geometry on such a scale. It is only on a very large scale (large compared to intergalactic distances) that the geometry would be significantly different from the Euclidean case (assuming the geometry was non-Euclidean).

Is it possible to have models of the universe which expand forever but in which space is finite, or alternatively, models in which the universe will stop expanding and eventually collapse but in which space is infinite? Both these possibilities can occur if one allows the so-called cosmological term in Einstein's equations. This term was originally introduced into his equations by Einstein himself, to enable him to obtain a static universe as a solution to his equations. Einstein was looking for a cosmological solution to his equations before Hubble's discovery of the expansion of the universe, and he was under the impression that the universe did not suffer any large-scale changes with time, that is, it was static. His original equations did not yield a static solution so he added the cosmological term to obtain a static solution. This term somewhat marred the simplicity of his original equations and Einstein later regretted introducing this term, for if he had relied on his original equations he might have been able to *predict* the expansion of the universe, or at any rate, to predict that the universe suffers large-scale changes with time. While the cosmological term yields a static solution, it can also lead to dynamic models of the universe, among which are models which expand forever but are finite in spatial extent, and also models which will stop expanding and collapse but are infinite in spatial extent. The dynamic models of the universe arising from the cosmological term are referred to as Lemaitre models after G. Lemaitre who first studied them.

How can we find out if the universe will expand forever or if it will stop expanding at some future time and start to contract? There are several interconnected ways of finding the answer. One way is to measure the present average density of the universe and compare it with a certain critical density. This comes about as follows. The recession of the galaxies is slowing down because of the attraction of different parts of the universe towards each other. Now the amount of gravitational force upon a typical galaxy that is slowing its velocity of recession depends on the density of matter; the larger the density at any given time, the larger this force of attraction. It turns out that if the density is above a certain critical density, the attractive force will be enough to halt the recession eventually and to pull the galaxies together. If the density is below the critical density, the attractive force is insufficient and the expansion will continue for ever. Of course the critical density itself, like the density of the universe, changes with time. The above somewhat crude argument can be made more precise and one can calculate the present value of the critical density. This value depends on the present value of the Hubble constant mentioned earlier. Recall that this is the number by which one has to multiply the distance of the galaxies (those that are not too near or not too far) to get their velocities away from us. In fact there is some uncertainty in the present value of the Hubble constant. The likely value of the Hubble constant is between 15 km/s to 30 km/s per million light years. That is, a galaxy which is 100 million light years away has a velocity away from us of 1500–3000 km/s. Cosmologists usually express the Hubble constant in terms of velocity of recession per megaparsec (million parsecs, that is 3.26 million light years). Using this unit the value of the Hubble constant falls between about 50 and 100. For a value of the Hubble constant given by 15 km/s per million light years, the critical density equals about 5×10^{-30} g/cm^3, or about three hydrogen atoms per thousand litres of space.

One can make a rough analogy of the situation just described with a missile thrown upwards from the Earth's surface. If the missile is thrown upwards with sufficient speed, it will slow down all the time but will nevertheless escape to infinity, whereas if it is thrown with insufficient speed, it will reach a maximum height and then return to the Earth. The situation in which the missile will escape to infinity corresponds for the universe to that in which the density is less than the critical, so that a typical galaxy has 'escape velocity'. The missile returning to the Earth corresponds to the situation above critical density and galaxies stopping to recede and collapsing. It can be shown that if the density is below critical, *all* galaxies will have escape velocity *all* the time and if the density is above critical, *all* galaxies will have less than escape velocity, so that they will all collapse. Thus expanding forever, or collapsing is a property of the universe as a whole.

The present density of visible matter and matter whose presence within galaxies can be deduced by various means is between one-tenth and one-fifth of the critical density as given by the value of 15 km/s per million light years for Hubble's constant. Thus it would appear that the universe is open, that is, it will expand forever. However, this is not certain as there may be matter in the intergalactic space whose presence has not been detected. Thus the subcritical nature of the present density is uncertain, and hence this method of finding whether the universe is closed or open is at present suggestive but not conclusive. The measurement of the density of the universe has been made more uncertain recently by the discovery that neutrinos may be massive. We proceed to explain the relevance of this important discovery.

As mentioned earlier, the cosmic background radiation indicates that there was a stage in the universe when matter and radiation were in equilibrium. In terms of the photon picture introduced in the last chapter we can think of the radiation consisting of numerous photons of various energies, being constantly absorbed and emitted by the charged particles, namely electrons, positrons and protons, which were present. There was equilibrium in the sense that as many photons in a certain energy range were absorbed by charged particles as were emitted in any given volume. Another way of saying this is to say that the 'mean free time' for a photon, that is, the average time for which a photon existed before being absorbed by a particle, was very short compared to the characteristic expansion time of the universe, that is, the time in which there was significant expansion of the universe. This equilibrium was possible in the early universe because electrons and protons were free due to the high temperature, that is, they were not bound together into atoms. This made them effective scatterers of the numerous photons that were present. A few hundred thousand years after the big bang, when the temperature dropped to about 3000 K, the electrons and protons and neutrons combined to make hydrogen and helium atoms. After this event there were no free charges present, or at any rate very few free charges left so that the photons were no longer constantly scattered and the universe became 'transparent'. From this time matter and radiation ceased to be in equilibrium. The radiation cooled as the universe expanded and eventually, at the present time, it has a temperature of about 3 K. The period during which electrons and protons and neutrons combined to make atoms is referred to as the 'recombination era'. After recombination the 'mean free time' for a photon became much larger than the characteristic expansion time of the universe. This is another way of saying that the universe became transparent, that is, a photon had to travel a long time (compared to the expansion time of the universe) before being scattered by some particle. One also says that after recombination matter and radiation became 'decoupled', that is they ceased to be coupled to each other in equilibrium.

There is reason to believe that in addition to the radiation there were present numerous neutrinos and antineutrinos in the very early universe which were in equilibrium with matter. When I say 'neutrinos' I mean all three types of neutrinos, namely the electron-neutrinos, the muon-neutrinos and the τ-neutrinos. I shall also mean antineutrinos when I say 'neutrinos'. In the first second or so after the big bang the 'mean free time' of the neutrinos was very short compared with the characteristic expansion time of the universe, so the neutrinos were in equilibrium with matter. A second or two after the big bang, when the temperature of the universe dropped below 10^{10} K and the density also decreased, the mean free time of the neutrinos (and antineutrinos) increased so that they began to behave as almost free particles. Thus the universe became transparent to neutrinos which went out of equilibrium with matter; matter and neutrinos were 'decoupled'. Just like the photons the neutrinos also cooled down and one can give arguments to show that neutrinos with a temperature of about 2 K should pervade the whole universe. This is far beyond present technology to detect.

Although neutrinos have energy, hitherto it has been assumed that they are massless. If indeed they are massless, the 'background' neutrinos would contribute a negligible amount to the present density of matter in the universe. But there has been a recent important indication that neutrinos may have a mass which is small compared to the mass of the lightest massive particle known so far, namely the electron. If indeed the neutrinos are massive, the 'background' neutrinos (which may be very numerous) may contribute a significant amount to the present average density of the universe. The neutrinos may even increase the density to above critical, so that the universe may be closed. There are, however, a number of uncertainties in this analysis. Firstly the precise mass of the neutrinos, if indeed they are massive (the three types of neutrino may have different masses), is not at present known. Secondly, the number density of the background neutrinos is also not precisely known. To make a crude analysis, let us assume that the number density of the background neutrinos is the same as the number density of the background photons. This latter number can be anything from about a 100 million to 20 000 million per baryon ('baryon' is a generic name given to protons, neutrons, and some more massive particles which we need not consider) in the universe. Let us settle for a figure of 1000 million photons per baryon. Now suppose that there are also 1000 million neutrinos per baryon in the universe. Since the mass of a baryon is approximately 1000 Mev, it is clear that if the average mass of the neutrinos is 10 ev the total mass of the neutrino will be about ten times the present density of matter, since the present density of matter is largely made up of baryons. In other words if our estimates are correct a mass of about 10 ev will cause the universe to be just closed. We are assuming here that the present density of the universe in baryons is one-tenth of the critical density. Of course very many uncertainties exist in the assumptions made in this analysis, so the value of 10 ev for a critical mass of the neutrino is very crude. In the near future, among the most important researches in cosmology will be making more precise the analysis which is given here very approximately, and also the experimental determination of the neutrino masses. Thus cosmology has become a truly experimental science in that experiments carried out in the laboratory can have a crucial effect on the theories of the structure of the universe.

It is possible in principle to find out if the universe is open or closed by measuring the rate at which the expansion of the universe is slowing down. The slowing down is measured by a certain parameter called the deceleration parameter. For the models of the universe we are mainly concerned with here, the deceleration parameter is related in a simple manner to the critical density; in fact in suitable units the deceleration parameter is just half the ratio of the actual density to the critical density. To measure the deceleration parameter one must look at very distant galaxies and correspondingly earlier times. This enables one to determine the rate of expansion at early times and compare it with the rate of expansion at later times. There is again considerable uncertainty in these observations because for the very distant galaxies it is difficult to decide if their dimness is due to distance or due to the fact that they are intrinsically dim, that is, they have lower absolute luminosities. One needs 'standard candles' of whose absolute luminosities one can be sure. This is difficult as the evolution of the universe and of galaxies may affect the absolute luminosities in an unknown manner.

One way in which absolute luminosities of distant galaxies can be affected is by so-called 'galactic cannibalism'. It was pointed out by J.P. Ostriker and S.D. Tremaine of Princeton University that larger galaxies may evolve not only because their individual stars evolve but also because some of them swallow their smaller neighbours. This happens by a process called 'dynamical friction'. Imagine a cluster of stars containing many stars. Consider now a much heavier star than those in the cluster coming into close collision with the outer members of this cluster. As a result of the collisions the heavier star slows down. Once it has slowed down the gravitational effect of the cluster as a whole on the heavy star becomes more effective and the heavy star 'gravitates' towards the centre of the cluster, that is, it gets swallowed up by the cluster. A similar phenomenon occurs in a satellite which goes round the Earth. The satellite feels the resisting motion (the 'viscous drag') of the thin atmosphere and gradually slows down. As its speed decreases the gravitational pull of the Earth becomes more effective and eventually the satellite is pulled down. (Actually there are two opposing effects because in the absence of friction the satellite nearer the centre of attraction has a higher speed.) Thus when a small galaxy is in orbit around a larger one, because of the interaction of the outer layers of the two galaxies the smaller one slows down and eventually spirals into the larger galaxy due to the gravitational pull of the latter (Fig. 5.3). This galactic cannibalism affects the absolute luminosities of 'standard candles' in the sense that absolute luminosities turn out to be lower in the very distant galaxies than was originally thought. This affects the measurement of the deceleration parameter. There are many such uncertainties and it will be some years before the deceleration parameter can be measured with some degree of confidence.

Another way to find out if the universe is open or closed is to determine the precise age of the universe and compare it with the so-called 'Hubble time'. Suppose there was *no* gravitational attraction of different parts of the universe towards each other. In this case the velocity of recession in the past would have been the same as at present. It can easily be shown that galaxies would have been 'on top of each other', that is, the big bang would have taken place at a time before now (called the Hubble time) given by the reciprocal of the Hubble constant in suitable units. For the value of 15 km/s per million light years for the Hubble constant, the Hubble time is about 20 billion years. In a closed universe the rate of expansion decreases more rapidly than in an open universe. Thus less time would have elapsed until the present time since the big bang in a closed universe than in an open universe. In fact one can show that if the age of the universe is less than two-thirds of the Hubble time then the universe is closed and if it is more than two-thirds of the Hubble time then the universe is open. However, present observational uncertainties both in the age of the universe and the Hubble time are too great to allow any definite conclusion to be drawn from this particular approach.

Fig. 5.3. A large galaxy may eventually swallow a small neighbour.

Most of the matter in the universe is in the form of hydrogen and helium. A hydrogen atom has one proton and one electron whereas a helium atom has two electrons around a nucleus consisting of two protons and two neutrons. Hydrogen comprises about 70–80% of all matter in the universe by weight, while helium comprises about 20–30% by weight. The heavier elements (hydrogen and helium are the lightest elements) make up a small percentage of all matter. Hydrogen is believed to be the primordial element out of which all the heavier elements were synthesized. One needs high temperatures to produce heavier nuclei from hydrogen nuclei, that is from protons. At high temperatures, in the presence of electrons, some of the protons change to neutrons by the inverse process to beta decay. That is, an electron and a proton combine to give a neutron and a neutrino. At high temperatures there is equilibrium between the protons and neutrons; the proportion of the neutrons depends on the temperature. There are three situations in the universe where high temperatures exist or have existed in the past: firstly, in the first few minutes after the big bang; secondly in the centre of very hot stars; and thirdly in supernova explosions (more about these in the next chapter). When a helium nucleus is created out of protons and neutrons, a certain amount of energy is released. If one estimates the total amount of energy released in the conversion of 20–30% of matter in the universe from hydrogen to helium, one finds that this is somewhat larger than the total amount of energy radiated by all stars in the universe since the creation of the stars some time after the big bang. This has led people to believe that helium was synthesized in the early stages of the universe rather than in stars. As regards the heavier elements, it was shown by E.E. Salpeter and E.M. Burbidge, G.F. Burbidge, W.A. Fowler and F. Hoyle that it is quite possible for most of the heavier element to be 'cooked' in the centre of hot stars and in supernova explosions. The process of 'cooking' heavier elements is called nucleosynthesis. The fact that heavier elements were synthesized in stars fits in with the fact that although it was possible to synthesize helium in the early universe, it was somewhat difficult to produce most of the heavier elements because the background radiation would have been at a very high temperature in the early universe and its intensity would have disintegrated any heavier nuclei as soon as they were formed. From the presence of so much hydrogen in the universe at present one can infer the presence of intense radiation in the first few minutes of the universe which prevented most of the hydrogen being converted into heavier elements. From the presence of this intense radiation in the first few minutes of the universe one can deduce the presence of a background radiation in the present universe. It is this chain of reasoning from which one can expect a cosmic background radiation. These calculations were done by P.J.E. Peebles and his collaborators at Princeton University, almost simultaneously with the discovery in 1965 of the cosmic background radiation by Penzias and Wilson. In fact long before that, in the late 1940s G. Gamow and his collaborators had predicted a background radiation with a temperature of 5 K. This work was largely ignored, partly because the theory of nucleosynthesis on which the prediction was based turned out to be erroneous in some respects.

Before the synthesis of helium in the early universe, protons and neutrons had to combine to make deuterium nuclei consisting of one proton and one neutron (deuterium is a constituent of heavy water, HDO). The deuterium nucleus was then synthesized into helium through the intermediate steps of tritium (consisting of one proton and two neutrons) and helium-three (consisting of two protons and one neutron). It is rather difficult to synthesize deuterium in stars because of its instability, so most of the deuterium that occurs in the universe is believed to be primordial, that is, that which was synthesized in the first few minutes of the universe. If the universe is closed, the present density of the universe is relatively large (greater than the critical density). If this is the case, the density of protons and neutrons in the phase of the universe in which nucleosynthesis occurred must have been correspondingly large. Thus a large density would have ensured that more of the deuterium that was formed would be converted into helium than would be the case for a smaller density. Thus one would expect a smaller abundance of deuterium at present if the universe is closed than otherwise. In principle, therefore, we can decide if the universe is open or closed by measuring accurately the present abundance of deuterium and comparing it with a certain critical density.

Although there are considerable uncertainties in all the above approaches, there are indications that the universe is open. Confirmation of this must await more accurate observations and better theories interpreting these observations.

6

Three ways for a star to die

In this chapter we shall discuss briefly how stars are born and how they evolve during their life, and then we shall consider in some detail how they eventually die, that is, reach the three final states of white dwarf, neutron star and black hole. We shall also discuss the phenomenon of supernova, a phenomenon which is relevant to the final state of some stars.

The precise manner in which stars are formed is not clearly understood. The region between the stars is not empty but consists of gas clouds, consisting mostly of hydrogen, and dust grains of various kinds. The material between the stars is not uniformly distributed in space but is spread in a patchy fashion. In most places the density of gas is very low, a typical density being 10^{-19} kg/m^3, that is about a hundred million (10^8) hydrogen atoms per cubic meter. Now the gravitational force between two portions of matter is inversely proportional to the square of the distance between them (for example, if the distance is doubled, the force becomes a quarter) and directly proportional to the product of the masses (if the masses of both portions are doubled, the force becomes four times as strong). Thus in a gas cloud, the higher the density of the gas, the stronger the gravitational attraction of different parts for each other.

Occasionally a cloud will become sufficiently dense and massive for gravitational attraction to draw it close together (this could conceivably happen also in the neighbourhood of a supernova explosion, as will be explained later). As the cloud begins to pull itself together, the density rises and so does the random motion of the particles forming the gas cloud. But this random motion is manifested as heat, so that the temperature of the gas rises and finally the star breaks into incandescence with a faint red glow. At this stage the star is shining by its gravitational attraction, that is, by the conversion of its gravitational potential energy into heat energy. At this stage the star is called a *protostar*. For example, the great nebula in Orion is believed to be a site of star formation. This nebula is situated about 1500 light years away and it contains a large number of protostars. The protostars emit radiation largely in the infra-red part of the electromagnetic spectrum. The detection of infra-red emitters in the Orion nebula (Fig. 6.1) supports the view that stars are being born there.

If gravitational energy was the only source of energy that the star possessed then it would not last very long. In fact it can be shown that for such a source of energy the Sun would last for only 20 million years or so whereas from geological studies it is known that the earth is approximately 4.5 billion years old, so that the Sun must be at least that old.

In fact stars get most of their energy from nuclear burning or nucleosynthesis, as follows. When the internal temperature of the star rises to a few million degrees, nuclear reactions occur at and near the centre of the star where the temperature is highest. At such high temperatures, the hydrogen atom (which consists of an electron and a proton) is stripped of the electron and one has essentially a gas of free electrons and protons. At such high temperatures there are frequent collisions between electrons and protons to produce neutrons and neutrinos. The neutron, in turn, decays in about ten minutes into a proton, electron and antineutrino, as mentioned earlier. Most of the neutrinos and antineutrinos escape from the star into outer space soon after they are formed, because they interact extremely weakly with other particles. From the mixture of protons and neutrons are formed helium nuclei consisting of two protons and two neutrons. A helium nucleus is also called an alpha particle. Once the helium nucleus is formed, the neutron in it becomes stable, that is, it does not decay any more. The helium nuclei are in fact formed out of neutrons and protons through the intermediate steps of deuterium, tritium and helium-three, as

Fig. 6.1. The Orion nebula, an important molecular cloud complex and star-formation site.

mentioned in Chapter 5. The overall effect is that four hydrogen atoms are converted into a helium atom. This process releases a great deal of energy. Where does this energy come from? The mass of a helium atom is slightly less than the combined masses of four hydrogen atoms. Thus when a helium atom is made from four hydrogen atoms the excess mass is converted into energy (in the form of electromagnetic radiation, that is, light, heat, etc.).This conversion takes place according to Einstein's formula $E=mc^2$ mentioned earlier. Here E is the energy released and m is the mass that is converted. If 1 g of mass is converted into energy, it produces approximately 10^{21} ergs of energy. The conversion of hydrogen into helium, known as fusion, is the process from which the hydrogen bomb gets its tremendous energy and which, hopefully, when controlled will provide energy for peaceful purposes. Thus one can say that stars get their energy by exploding millions of hydrogen bombs every second.

The heat and radiation produced firstly by the conversion of gravitational potential energy and secondly by nuclear burning causes the material in the star to have an outward pressure which temporarily balances the gravitational pull towards the centre of the star felt by a typical portion of the star. The star then acquires a 'metastable' (approximately stable) state in which it can last for a few million years to many billions of years depending on various circumstances. During this period the star is said to be on the *main sequence*. This period is the longest in the star's life. Usually in a more massive star the rate of conversion of mass into energy is higher than in a star of lower mass (the Sun is a comparatively low mass star). Thus the Sun produces energy at the rate of about 2 ergs/g/s, whereas a more massive and luminous star than the Sun may produce energy at the rate of 2000 ergs/g/s. Thus for many more massive stars the metastable or main sequence stage of nuclear burning may last for only a few tens of millions of years, whereas for the Sun this stage has lasted for at least 4.5 billion years and may last for many billions of years more.

The nuclear burning in the star proceeds not only by the conversion of hydrogen into helium, but also by the conversion of helium into heavier elements. But the next stage usually requires a higher temperature. When all the hydrogen has been converted into helium in the central portion or core of the star, there is not enough energy left in the core to withstand the pull of gravity inwards, so the core begins to contract. As it contracts it gets heated up by the conversion of gravitational energy into heat energy. The temperature then rises and reaches a point at which the next stage of nuclear burning can proceed. In these reactions helium gets converted into carbon, then carbon to oxygen, then to neon and so on. At each stage of the series, the nuclear reaction takes place by the addition of a helium nucleus or alpha particle (that is, two protons and two neutrons). The same process that happened after hydrogen was exhausted repeats itself after the formation of each species is completed. That is, the core shrinks due to lack of outward pressure, then heats up as it shrinks, and triggers off the next series of nuclear reactions. The nuclear reactions at each stage temporarily halt the shrinkage of the core. At higher temperatures bigger nuclei can be formed, for example in carbon burning two carbon nuclei are fused to give a magnesium nucleus and, as usual, the fusion releases a great deal of energy. There is, however, a limit to this process because beyond a certain mass the nucleus is not very stable and so higher temperatures do not necessarily produce heavier nuclei with release of energy. The limit is reached in the iron group of nuclei, namely, iron, cobalt and nickel, each of which possesses a total of 56 neutrons and protons.

While these reactions are taking place in the core, which progressively contracts and becomes hotter, the outer envelope usually expands and cools (see Fig. 6.2). The precise reasons for the expansion of the outer layers cannot easily be explained physically, although this is the result that is often derived when the equations describing the star are solved. To an outside observer the star appears to become larger in size and to

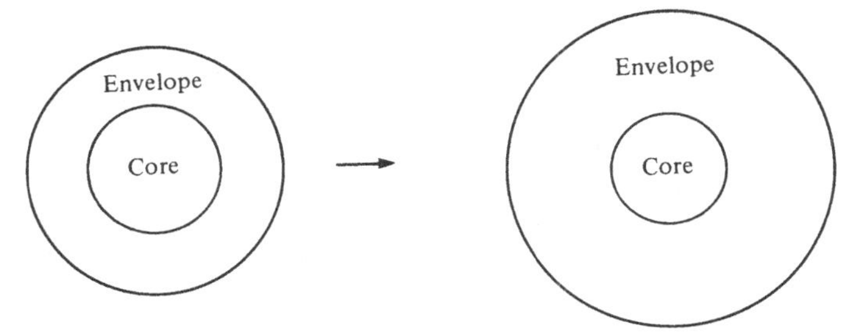

Fig. 6.2. The core of a low-mass star such as the sun shrinks and its envelope expands to produce a 'red giant'. The figure is not drawn to scale.

become redder in appearance (the visible effect of cooling). We then get a so-called 'red giant'. This, at any rate, happens to many low mass stars. The Sun, for example, is likely eventually to become a red giant and expand to the orbit of Mercury, Venus or even the Earth.

The exact evolutionary process for stars of various types is still not clearly understood in all the details. We are, however, concerned mainly with the final stages, which are better understood. Eventually the nuclear fuel of the star gets exhausted as most of the hydrogen gets converted into helium and helium into heavier elements. There is then not enough energy produced in the star to balance the inward force due to gravitational attraction, and the star begins to contract. As it contracts the matter in the star become more and more dense. At some stage the density is so high that the electrons in the material of the star get stripped off and the star becomes a collection of nuclei of helium and heavier elements, with all the electrons being separated from the nuclei, running around freely in the star. The electrons provide a certain outward pressure which can be explained as follows.

All elementary particles in the universe can be divided into two categories, called fermions and bosons (named, respectively, after the Italian physicist Enrico Fermi (1901–1954) and the Indian physicist Satyendra Nath Bose (1894–1974)) which behave entirely differently under conditions of high density or very low temperatures. Examples of fermions are electrons, protons and neutrons, and an example of bosons is the photon. Fermions obey the so-called Fermi–Dirac statistics, which is a theory of the behaviour of large aggregates of particles propounded by Fermi and Dirac. Bosons obey the Bose–Einstein statistics, formulated by Bose and Einstein. It is one of the consequences of quantum theory and relativity that bosons have integral intrinsic spins, that is, they have spins 0, $1\hbar$, $2\hbar$, etc., whereas fermions have half-odd-integral spins $\frac{1}{2}\hbar$, $\frac{3}{2}\hbar$, etc. Thus electrons, protons and neutrons have spin $\frac{1}{2}\hbar$ and so they are fermions, whereas the photon has spin $1\hbar$ so it is a boson. Fermions have the property that no two identical fermions can occupy the same state (this is known as the Pauli Exclusion

Principle after Pauli, mentioned earlier, who first enunciated this principle). What exactly does one mean by a 'state'? It is difficult to define this quantum mechanical concept in general but two examples may elucidate it. According to the laws of quantum mechanics, in an atom the electrons can have only certain discrete orbits around the nucleus. Now the electron has an intrinsic spin which can have one of two orientations. In an atom the state of an electron is specified by its orbit together with an orientation of spin. Thus according to the Pauli Exclusion Principle at most two electrons can occupy a certain orbit, the two electrons having the two possible orientations of spin. Another example of a state arises when considering an electron gas confined to a box. Each state in this case is specified by the energy of the electron, its momentum components (the latter in fact determines the energy) and the spin. Again according to quantum mechanics the energy and momentum of particles confined to a finite box can take up only certain discrete values. Since only one electron can occupy each state there develops a resistance to compression or an outward pressure if too many electrons are confined to a small volume. Thus it is the fermionic properties of electrons that cause the outward pressure mentioned at the end of the previous paragraph. This pressure is referred to as 'Fermi pressure' or 'electron degeneracy pressure'.

The amount of gravitational force upon a typical star towards its centre depends on the total mass of the star. The higher the mass, the stronger the force, in general. The outward Fermi pressure of the electrons in the star can balance the gravitational force inwards only if the mass of the star is less than 1.4 times the mass of the Sun. This limit to the mass of the star is known as the Chandrasekhar limit after the Indian astrophysicist (now a USA citizen) S. Chandrasekhar, who discovered this limit in about 1935. If the mass of the star is less than $1.4M_{\odot}$ ($M_{\odot}$ stands for the mass of the Sun) then the gravitational force balances the electron Fermi pressure and we get the so-called white dwarf star. These stars have masses of the order of the mass of the Sun but are about as big as the Earth. For example, the white dwarf 40 Eridani B has a mass of about $0.44M_{\odot}$ but has a radius only 1.5% of that of the Sun. The matter in these stars is very dense, anything from ten thousand times to about a million or more times as dense as water. Such stars eventually cool off and become less and less luminous to become what are sometimes called 'black dwarfs'. Although they are rather faint objects, many white dwarfs have been discovered. One of the first white dwarfs to be discovered is a star which forms a binary system (two stars revolving about each other) with the relatively bright star named Sirius and is known as the companion of Sirius. Sirius and its companion are sometimes known as Sirius A and Sirius B.

If the mass of the star is greater than $1.4M_{\odot}$ the gravitational force inwards overwhelms the Fermi pressure of the electron outwards and the star continues to contract further. The material of the star ultimately becomes so dense, especially near the centre, that the electrons are squeezed into the protons of the nuclei to form neutrons (and neutrinos which escape) and different nuclei merge into one another to form essentially a single giant nucleus of neutrons comprising a substantial part of the star. As in an ordinary nucleus, in this giant nucleus neutrons are stable and do not decay as a free neutron would. Neutrons being fermions, they now exert a neutron Fermi pressure just as electrons exerted electron Fermi pressure. In addition to the Fermi pressure of the neutrons there may be some additional pressure present because of the nuclear forces between neutrons (such forces also exist between neutrons and protons, also between protons and protons, and hold the nucleus together). This is not completely understood at present because the very short-range behaviour of nuclear forces is not known. The pressure created by the neutrons in this manner can balance the force of gravity only if the mass of the star is less than about three times the mass of the Sun, that is, less than $3M_{\odot}$. This limit of $3M_{\odot}$ is only approximate. To determine this limit accurately one has to know the so-called 'equation of state' of the material of the giant nucleus consisting of neutrons. The equation of state is an equation which enables one to determine the pressure from the density of the material and vice versa. The equation of state for nuclear material is not known because of the lack of knowledge of nuclear forces at short range. However, one can make some general assumptions about this equation of state which should be valid because they are based on general physical principles which are known to be valid. By making such general assumptions about the equation of state one arrives at a mass of about $3M_{\odot}$. Future calculations should make this mass limit more precise.

When the mass is less than about $3M_{\odot}$ we get a so-called neutron star, in which a large portion of the inner core of the star consists of a giant nucleus of neutrons. There is a class of astronomical objects known as 'pulsars' which was discovered by the Cambridge (England) astronomer A. Hewish and his collaborators in 1967 (see Fig. 6.3). These objects send out rapid pulses of radiation at very regular intervals of the order of a second. It is now generally believed that these pulsars are in fact rapidly-rotating neutron stars. The original suggestion that they are rotating neutron stars was made by T. Gold of Cornell University. Over 300 of these pulsars have been discovered since 1967. The periods of the pulsars range from a few milliseconds (a thousandth of a second) to about four seconds. These periods show slight lengthening over the years. This is consistent with the fact that rotating neutron stars are dissipating energy to the surrounding medium and consequently slowing down. The material of the neutron star is extremely dense, from about 10^{13} to 10^{15} g/cm^3. This is the density of the nucleus of an atom. A small spoonful of this material would weigh several million tons on the Earth. The mass of a neutron star is of the order of a solar mass but its radius is only about 10 or 20 km. The discovery of pulsars has provided modern science with an excellent 'laboratory' in which to explore all sorts of exotic phenomena which cannot be reproduced in a terrestrial laboratory.

As mentioned earlier, the neutron was discovered by Chadwick in 1932 (Fig. 6.4). It is alleged that the very day the news of this discovery reached Copenhagen the Soviet physicist Lev Davidovich Landau (1908–1968) discussed the theoretical implications of this news with colleagues. There are anecdotes to the effect that it was Landau who first suggested

Fig. 6.3. Two of the antennae of the One-Mile Telescope at the Mullard Radio Astronomy Observatory, Cambridge. Another telescope was used for the discovery of pulsars; this telescope obtained precise positions of the pulsars.

that there may exist cold dense stars composed primarily of neutrons. Quite independently, W. Baade and F. Zwicky in 1934 proposed the idea of neutron stars. They also put forward the suggestion that such stars may be formed in supernova explosions. Zwicky, in 1939, also suggested that the vast energy

Fig. 6.4. The Cavendish Laboratory in Free School Lane, Cambridge, where J. Chadwick discovered the neutron.

of the supernova explosion might be derived from the creation of a tiny neutron star. Zwicky is also credited with the idea that there might be a neutron star in the Crab nebula (Fig. 6.5).

A neutron star can have a mass of less than $1.4M_{\odot}$, although for this range of mass the white dwarf configuration is possible. Sometimes the central portion of a star experiences a sudden

Fig. 6.5. The Crab nebula, remnant of the supernova sighted in AD 1052.

inward force or an implosion due to an explosion taking place in the outer layers of the star (see Fig. 6.6). This inward pressure causes the Fermi pressure of the electrons to become insufficient to withhold gravity and electrons are squeezed into protons to become neutrons, etc., and the inner core becomes a neutron star while the outer layers are dispersed into the surrounding space. The inner core may have a mass of less than $1.4M_{\odot}$, in which case we get a neutron star of mass less than $1.4M_{\odot}$. In fact a neutron star can have a mass as low as $0.2M_{\odot}$. For masses more than $3M_{\odot}$ the remnant is a black hole, as explained later. This kind of explosive phenomenon is found to occur in the so-called 'supernova' mentioned earlier in which a star is observed to suddenly increase its brightness by a factor

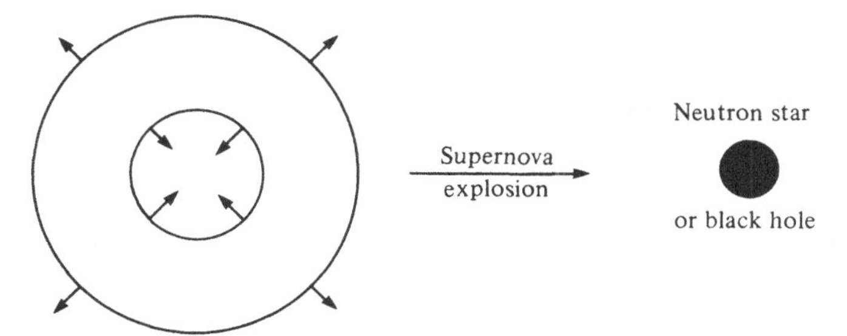

Fig. 6.6. In a supernova explosion the core implodes while the outer envelope explodes to create a remnant which can be a neutron star or a black hole.

of tens of billions within a matter of days. One such supernova was seen in AD 1054 by Chinese astronomers; the remains of this explosion can still be seen in the Crab Nebula, in the midst of which a pulsar has been discovered. This discovery supports the theory that neutron stars can be created in supernova explosions. But it may not be the case that the formation of a neutron star is always associated with supernova explosions. Other examples of supernova in our Galaxy are the ones of AD 1572 (Tycho's Nova), 1604 (Kepler's star) and 1843 (Eta Carinae). Supernovae in other galaxies have also been seen. One of the brightest to be seen this century was the supernova of 1937, which was a hundred times brighter than the dwarf spiral galaxy IC4182 in which it appeared.

There are various kinds of supernovae; I have described one possible sequence of events. The exact causes of supernovae are not yet clearly understood. Some kind of runaway thermonuclear reactions in the star could be partly responsible for these. The actual explosion could be caused by shock waves produced at the time of the formation of a neutron star. A shock wave is a region of discontinuity (sudden change) in a medium in which sharp changes occur in the properties of the medium such as its pressure and density; shock waves often occur when the velocity of the material of the medium exceeds the speed of sound. A supernova explosion could also be triggered off by the neutrinos leaving the core during the formation of a neutron star (recall that when an electron and a proton combine to form a neutron, a neutrino is given off). The shock wave or the flux of neutrinos produces high temperatures as it travels outwards through the outer envelope of the star. The outermost envelope is probably hydrogen in which the temperature has never been high enough for nuclear reactions to take place. The next layer probably consists of helium, and then nuclei of heavier elements. As the shock wave travels through these media, the high temperature sets off nuclear reactions. This process is referred to as *explosive nucleosynthesis*, in which the temperature is very high temporarily and causes runaway nuclear reactions. The precise details are still not known and are currently under investigation by astronomers.

There is some evidence that the shock waves which are sent out through the star and eventually through the surrounding medium of gas in a supernova explosion help in the formation of new stars. Recall that the density of interstellar matter is usually too low for gravity to draw the cloud together to form a protostar; this can happen only if the density becomes sufficiently high. As the shock wave from a supernova travels through the surrounding medium, it causes the density in some regions to become high enough for star formation. There is some evidence that some new stars may be born in this manner.

It is known that for a star to undergo a supernova explosion it must have a mass of at least six times the mass of the Sun. Thus the Sun is very unlikely to become a supernova. The Sun, after becoming a red giant, will probably settle down to become a white dwarf.

When the mass of the core is greater than about $3M_{\odot}$, even the neutron Fermi pressure and forces of nuclear interaction are not sufficient to withstand the force of gravity. In this case we get a 'black hole', which will be discussed in detail in the next chapter.

The three final states of white dwarf, neutron star or black hole occur for stars with masses which are not too small compared with the mass of the Sun. For smaller masses like the Earth or a piece of rock, gravity can be balanced indefinitely by the ordinary pressure that matter exerts in resisting compres-

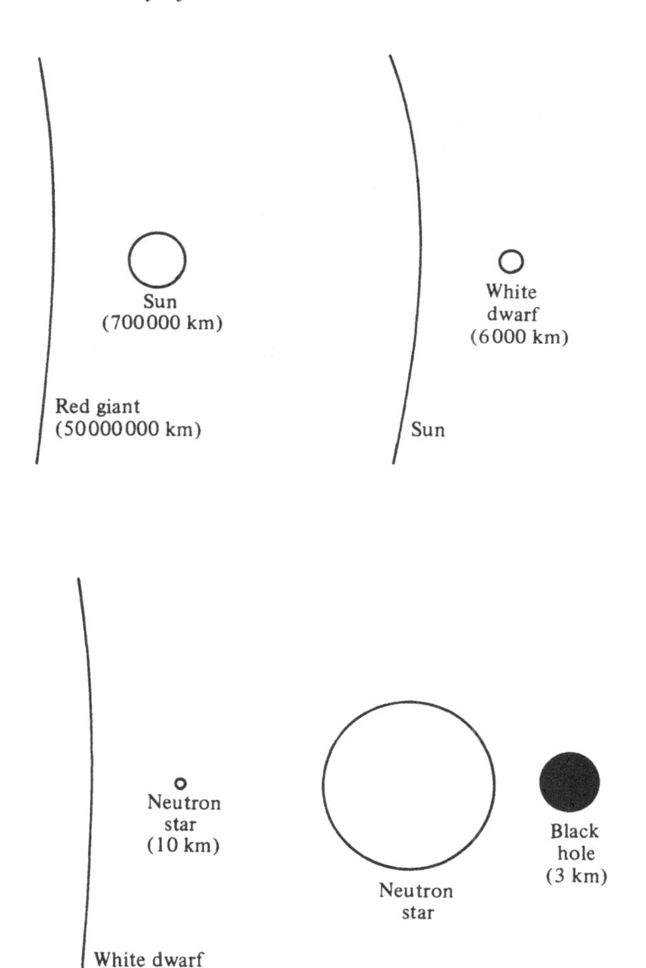

Fig. 6.7. How far gravity can shrink a star depends on its initial mass. Here are shown the relative sizes of a one-solar-mass star after attaining the different end states described in the text. The sun is presently 700 000 km in radius. After the sun has become a red giant in about five billion years from now, at the end of its hydrogen-burning phase, and undergone helium burning, its core should collapse, forming a white dwarf. Heavier stars may contract hundred of times further, to neutron stars or even black holes.

sion. In fact in a piece of rock or the objects of our daily experience gravity plays scarcely any part in their structure. This structure is determined essentially by electrical forces between atoms. In the Earth or the Moon, although gravity plays a part in the structure (for example, in making the Earth and the Moon approximately spherical), the gravitational pull inwards can be indefinitely balanced with little change in size and shape no matter how cold these bodies become. It is only with a mass approximately that of the planet Jupiter (whose mass is of the order of a thousand times that of the earth) that gravity begins to play a dominant role (Fig. 6.7).

7

Black holes and quasars

As mentioned in the last chapter, when the mass of the star is greater than about three times the mass of the Sun, even the neutron Fermi pressure and other outward forces exerted by neutrons are not sufficient to withstand the force of gravity. There are no known forces in nature that can balance the force of gravity under these circumstances, and the star collapses and collapses until it reaches a very small volume with very high density. The precise nature of the final form that matter takes in this case is not known. What is indicated by Einstein's theory of gravitation is that when all the matter in the star goes within a certain small volume, no further communication is possible with the matter inside this volume, since any rays of light (which are the fastest possible signals) leaving it are pulled back towards the central region by the strength of gravity. The star then becomes a 'black hole'. It is called 'black' because no radiation of any kind comes from within it. If the star has no rotation initially, the black hole quickly settles down to a spherical shape, the radius of the hole (from within which nothing can come out) being dependent on the total mass. For a mass M, the radius of the black hole, known as the Schwarzschild radius (after the German astronomer Karl Schwarzschild (1873–1916), who first found the solution of Einstein's equations in 1916 corresponding to a black hole) is $2GM/c^2$, where G is the Newtonian gravitational constant and c is the velocity of light. If M is the mass of the Sun, which is about 2×10^{33} g, then the Schwarzschild radius is about 2.95 km. If M is the mass of the Earth, which is about 5.98×10^{27} g, the Schwarzschild radius is about 0.89 cm. Strictly speaking, one should describe the size of a black hole in term of its circumference rather than its radius. This is because the radius is not directly measurable since one cannot go inside the black hole and come out.

Although a star of mass of greater than about $3M_{\odot}$ must eventually become a black hole, it is possible to have black holes of mass less than $3M_{\odot}$. In this case the black hole is not produced simply by the strength of the gravitational pull inwards, but some external agent must apply additional force. For astronomical bodies this external force comes from supernova explosions. As mentioned in the last chapter, for a star to become a supernova its mass has to be greater than about $6M_{\odot}$. However, in a supernova explosion it is the core which implodes and this core can have a much smaller mass. After the supernova explosion the core becomes a neutron star, or, if the pressure inwards is sufficiently strong, it can also become a black hole. Thus black holes of the mass of the Sun or less are also possible, but these need external agents such as supernova explosions. For a mass of about ten solar masses, say, the black hole can be produced by the strength of the gravitational force alone without a supernova explosion having to press it inwards. The earth, for example, has to be squeezed by external forces to less than a circumference of about 5.58 cm before it can become a black hole.

As early as 1798 the French mathematician and astronomer Pierre Simon Laplace (1749–1827) wrote, 'A luminous star, of the same density as the Earth, and whose diameter should be two hundred and fifty times larger than that of the sun, would not, in consequence of its attraction, allow any of its rays to arrive at us; it is therefore possible that the largest luminous bodies in the universe may, through this cause, be invisible'. It was a remarkable insight on Laplace's part to realize that gravity could be strong enough in some circumstances not to allow light to escape. In this sense he anticipated the black hole. However, the star that Laplace had in mind could not remain in equilibrium; it would collapse and give rise to a black hole.

A black hole is a simple object in the sense that it is described completely by a single parameter, namely, the mass if it is a non-rotating black hole. A rotating black hole is one which is formed out of a rotating star. In this case the black hole is not spherical in shape but flattened at the poles. A rotating black hole, which is represented by a solution of Einstein's equations found by R.P. Kerr, is completely described by two parameters, namely its mass and angular momentum (which is related to its speed of rotation). The surface around a black hole out of which no signals can emanate is called the *event horizon* or simply horizon. Exactly how simple black holes must be has been discovered by B. Carter, S.W. Hawking, W. Israel and D.C. Robinson. They have shown that when a black hole first forms, the horizon may have an irregular shape and may be wildly vibrating. Within a fraction of a second, however, the horizon should settle down to a unique smooth shape. This shape is absolutely spherical if there is no rotation initially. It is flattened at the poles if there is rotation present initially; the amount of flattening depends on the speed of rotation. The black hole retains no memory of what sort of matter has gone into it; all properties of the matter have been obliterated except its total mass and the total angular momentum. Even the information whether the matter consisted of particles or antiparticles (that is, whether it was matter or antimatter) is lost. As we shall see in a later chapter, there is an additional property of matter that the black hole retains, that is the total electric charge. But this quantity is usually zero for the matter that forms black holes from stars, since stellar material usually has no net total charge, that is, it is electrically neutral.

The gravitational attraction that the black hole provides for light can be illustrated by Fig. 7.1. Here the large circle is the event horizon of a black hole. In the centre, about the nature of which very little is known, is the so-called singularity at which space time may have weird properties. In this figure the light signal originating at any one of the points shown occupies at the next instant a spherical surface (small circles) known as its wave front. At large distances from the black hole the point lies at the centre of this surface. At smaller distances the wave front is displaced towards the black hole because of its gravitational attraction. At a point on the event horizon the spherical wave

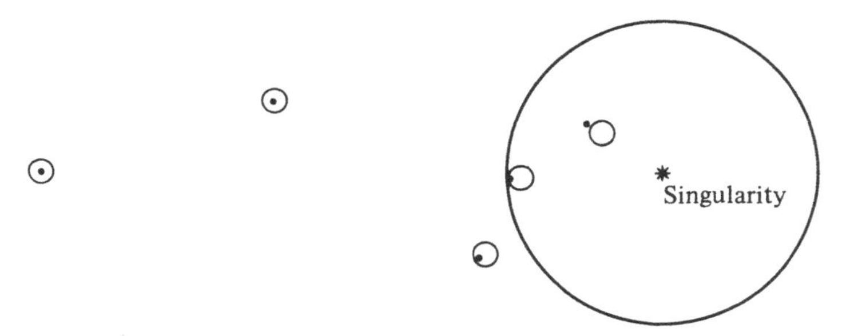

Fig. 7.1. The large sphere indicates the event horizon of the black hole. The small spheres are the positions of the wave fronts of light emitted at the nearby points an instant before. These wave fronts are dragged towards the black hole as the emitting point comes nearer the centre of the black hole.

front touches the horizon internally and never escapes it. At interior points the wave front leaves the emitting point altogether as soon as it is emitted, and is sucked in towards the singularity.

It is not possible to observe a black hole directly because it gives off no radiation. However, it still exerts a gravitational force on surrounding bodies and it may be detected indirectly from its effects on nearby bodies, for example, if it happens to be revolving around another star forming a binary system. With this idea in mind in 1964 two Soviet astrophysicists Ya.B. Zel'dovich and O.Kh. Guseynov made a systematic search through spectroscopic binaries. These are twin stars which appear as a single star even to the most powerful telescopes, but if one studies their spectra, one finds blue shifts and red shifts, indicating that there are two stars, one of which is approaching the earth and the other moving away. This shows that there are two stars revolving about each other. Sometimes even spectroscopic analysis does not reveal two stars directly but from the motion of one of them (showing blue shifts and red shifts alternately over certain periods) one can deduce that a star is revolving around an unseen companion. Such a binary system is a possible place to look for a black hole. Of course there may be an alternative explanation for the companion to be unseen; it may simply be a very dim star. It is possible by studying the periods of the binaries and by various other means to determine the masses of the stars. A candidate for a black hole must have a mass of greater than three solar masses.

How can one make sure that the unseen companion in a binary system is indeed a black hole? A black hole in a close binary system might pull off gas and other material from its companion and there might be radiation from this material just before it enters the horizon. Since the gravitational field is very strong near the horizon, the material falling into the black hole acquires high velocity and acceleration and so gives off high-energy electromagnetic radiation, probably mostly in the X-ray band. The radiation comes from the matter *before* it crosses the horizon; any radiation given off after the matter enters the horizon is sucked in by the black hole. A search for X-rays given off by binary systems cannot be carried on by ground-based instruments because the atmosphere is opaque to X-rays. An X-ray telescope was launched jointly by the USA and Italy aboard the Uhuru satellite in December 1970. By 1972 enough data were available to compile a list of over a hundred X-ray sources. None of the list compiled by Zel'dovich and Guseynov was recorded by Uhuru to be an X-ray source (this does not necessarily mean that these binaries do not have black holes) but the Uhuru list contained six binary X-ray sources which had previously not been recognized as binary systems. Two of these systems turned out to harbour neutron stars because the X-rays from these two came at regular intervals showing the presence of pulsars. Such regular pulses cannot come from black holes because the latter cannot harbour magnetic fields round them in the same way that rotating neutron stars can. The regular pulses come from the interaction of the rotatory motion of the neutron star and its magnetic field.

The four remaining X-ray binary sources have been under intense investigation for possible black holes. The most promising candidate for a black hole is the X-ray binary source Cygnus X-1. There is an unseen companion which is the source of the X-ray and which has a mass of about eight or ten solar masses and which is rather compact. Thus the mass and size is consistent with the companion's being a black hole. Alternative models for Cygnus X-1 have been proposed, but these do not appear to be satisfactory. However, a great deal of work remains to be done to establish definitely whether or not the unseen companion in Cygnus X-1 is a black hole. Both theorists and experimentalists are actively searching for ways in which this question can be settled. Should it turn out that Cygnus X-1 indeed has a black hole, it will be one of the most important discoveries of this century.

Another class of objects which is of great interest as regards cosmology is the so-called quasars. There is some controversy about the precise nature of these objects. Consider the quasars called 3C48 and 3C273. The prefix 3C means that these appear in the *Third Cambridge Catalogue* (as the 48th and 273rd objects, respectively), which is a catalogue of radio sources published by Cambridge (England) astronomers in 1959. The term 'radio sources' means that these sources in the sky emit strong enough radiation in the radio band to be detected by radio telescopes on the earth. As explained earlier, stars and galaxies emit electromagnetic waves of all wavelengths. Some give off more energy in the visible part of the spectrum, that is, light, and therefore are optically bright, while others may give off more radio waves and may be powerful radio-emitters. The sources 3C48 and 3C273 were also identified through optical telescopes and they appeared to be stars. However, they showed peculiar emission lines. We have explained earlier how dark lines appear in the spectrum of stars and galaxies due to the absorption of light by cooler gas in the outer layers of the star or galaxy. Emission lines, on the other hand, are bright lines in the spectrum of the object caused by hot gas or material within the object. Emission lines can show red shifts (or blue shifts) just as the absorption (dark) lines. The sources 3C48 and 3C273 were at first thought to have peculiar emission lines because it was thought that they were stars within our Galaxy. Eventually it turned out that the emission lines in these objects were familiar lines which had been red shifted by the equivalent of $z = 0.367$ and $z = 0.158$, respectively. According to the Hubble interpretation, these were roughly at a distance of 5 billion and 3 billion light years away, respectively. Since the discovery of these two quasars in the early 1960s, many such quasars have been discovered with high red shifts and star-like appearance. For example, the quasar 3C9 has a red shift of $z = 2.012$; if this red shift is due to the expansion of the universe according to Hubble's Law it should be at a distance of over 10 billion light years. An even higher red shift of $z = 3.53$ has been discovered.

One feature of the quasars which adds to the puzzle is that many of them show enormous variation in their brightness over periods of weeks or months. At the peak of their brightness they shine with the brightness equivalent to ten thousand times the brightness of an ordinary galaxy. Now the fact that the variation in the brightness takes place within a matter of weeks or months means that the region from within which the radiation is emerging must be a few light weeks or light months across. This is because the region as a whole must be producing energy and different parts of it cannot be affecting each other at a rate faster than the speed of light. Compare this region with, for example, the size of our Galaxy which is about 80 thousand light years across. Thus the energy output from only a few light weeks or light months equals the energy of ten thousand galaxies! How this enormous energy is produced from such a small region has been a great puzzle. One possible answer is that at the centre of a quasar is a large black hole of perhaps a hundred million solar masses. The Schwarzschild radius of such a black hole is about 16 light minutes, that is, about twice the distance from the Sun to the Earth (this is about 2 astronomical units). This black hole swallows up stars from outside its horizon, and in the process of being swallowed up the star is disrupted by the 'tidal' gravitational force of the black hole. The tidal gravitational force arises because the gravitational force from the black hole is different at different distances from the centre so parts of the star nearer the black hole feel a stronger force than parts further from the black hole. This tends to elongate the star and eventually to disrupt it. In the process of such disruption a great deal of energy is given off. Such a process can account for the energy output of quasars provided this enormous output does not last for more than a few tens of millions of years at most. There is indication that this is so, that is, quasars do not last for more than a few tens of millions of years. The fact that this 'black hole model' is a successful one for quasars could be regarded as indirect evidence for the existence of black holes.

There is a small but vocal minority of astronomers which maintains that the red shifts of quasars are due to other causes than the expansion of the universe and therefore they are not at enormous distances as indicated by the red shift. For example, they could acquire a high red shift if the quasars were ejected at high velocities from some other galaxies, in particular from the centre of our Galaxy. However, the centre of our Galaxy does not seem to have had such violent events. If the quasars are ejected by other galaxies, it would seem that we should see some blue shifts in addition to the red shifts, as some quasars may be ejected towards us. The eminent astronomer F. Hoyle, who is of the opinion that quasars are not at large distances, recently claims to have found an explanation for the fact that we do not see blue shifts (on the assumption that quasars are ejected from galaxies). On the observational side the view that quasars are not 'cosmological' (that is, at large distances) is championed by H.C. Arp of Hale Observatories in California. Arp has found several instances of galaxies which appear close to each other in a photographic plate but have different red shifts (Fig. 7.2). For example, he found a chain of five galaxies, four of which have a red shift of $z = 0.05$ while the fifth has a much higher red shift of $z = 0.12$. The claim is that the five galaxies are actually near to each other and that the higher red shift of the fifth one is due to reasons other than the expansion of the universe. However, the majority view is that Arp's examples arise from chance alignment of the galaxies along the line of sight, that is, the galaxy with the higher red shift is actually much further away but only appears to be near the others in the photograph. In the 'local' view of the quasars the energy problem is not so great. That is, quasars have enormous output of energy only if they are assumed to be at large distances. If they are regarded as local then it is not necessary to

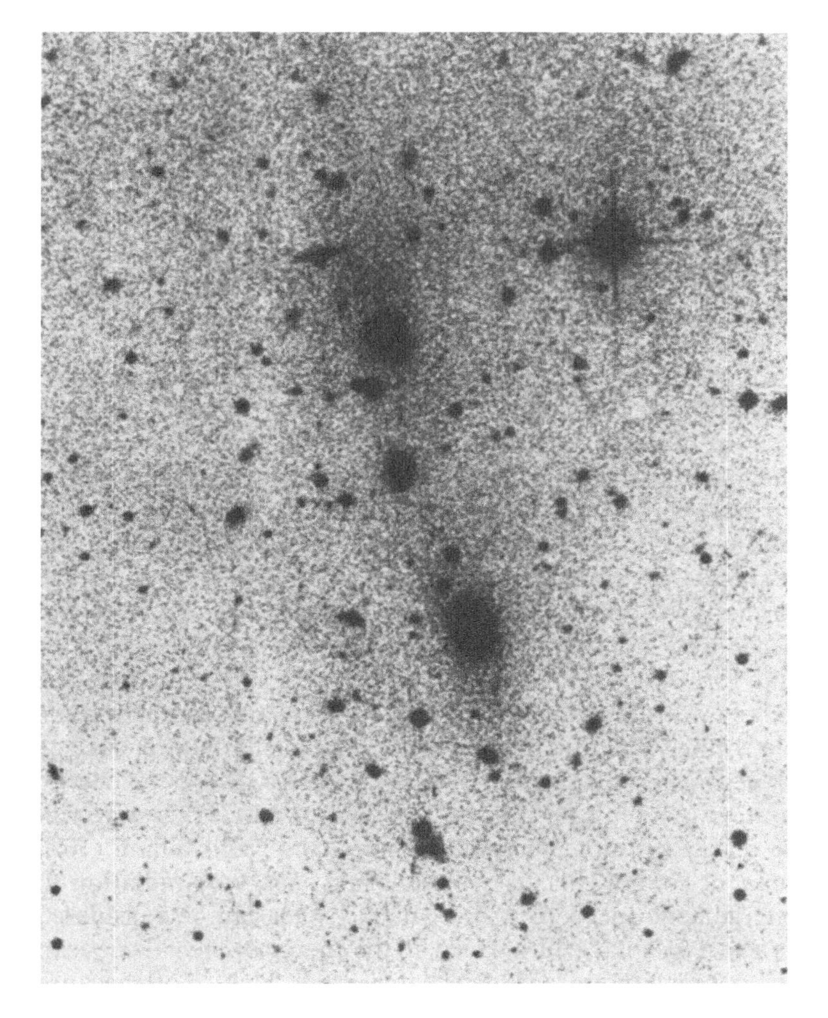

Fig. 7.2. A chain of galaxies, Arp 324. (Negative photograph.)

invoke a large black hole and their energy output can be understood through more conventional processes. The question whether quasars are local or cosmological is a very important one which is actively under consideration by astronomers. The majority view is that they are cosmological but much remains to be done to settle the matter finally.

8

Galactic and supergalactic black holes

From the last two chapters it is evident that all stars in a typical galaxy will eventually be reduced to white dwarfs, neutron stars or black holes. There will be formation of new stars from the interstellar gas but eventually most of this gas will be used up in making stars which will eventually die. The remaining gas will be too thinly dispersed and cold to make new stars. The remnants of supernova explosions could also lead to the formation of new stars but finally these remnants would become too rich in heavy elements for the normal process of star formation to take place. Thus given sufficient time, the galaxy will simply consist of cold white dwarfs, neutron stars, black holes and other forms of cold interstellar matter such as planets, asteroids, meteors, rocks, dust, etc. From the energy content of a typical galaxy, it can be shown that this stage will be reached in not much more than a thousand billion (10^{12}) years or so. All galaxies will be losing energy by radiation to intergalactic space. The intergalactic space can be considered as a vast receptacle into which all the energy of the galaxies can be poured without raising its temperature. This is both because the empty space between galaxies increases as the universe expands and because the radiation given off by galaxies gets red shifted and becomes weaker. Some of the energy from the radiation goes into the work done in the expansion of the universe. In 10^{12} years or so there will be scarcely any radiation coming out of the galaxies. The sky will then look pitch black to human type eyes everywhere in the universe, except for some occasional flashes of light and radiation from the centre galaxies, somewhat later on, as we shall see later in this chapter.

In this chapter and subsequent ones (until chapter 11) we shall assume that the universe is open, that is, it will expand forever. This implies that an infinite time is available in the future.

After 10^{12} years or so the cold white dwarfs, neutron stars, black holes and other forms of cold interstellar matter will still form a galaxy, that is, they will still be bound together in their mutual gravitational field. The long-term evolution of such a system is very difficult to predict exactly or even approximately. This is one of the unsolved problems of astronomy. However, one can carry out a very rough analysis and attempt to make some plausible conjectures. Let us begin with the simplest case of this problem, namely, two bodies going round each other in their mutual gravitational field, that is, under the force of their mutual gravitational attraction.

Suppose we have two neutron stars going round each other. According to Newtonian theory of gravitation, they will each describe an elliptical orbit with the centre of mass of the two bodies at one of the foci of the ellipse. Even this is not strictly accurate when one considers the very long-term behaviour of such a system. Neutron stars are extended bodies, that is, they are not concentrated at points. The orbits of two bodies going round each other can be calculated exactly according to Newton's theory of gravitation only if the two bodies are considered as 'point' bodies with no extension. In such a case the result is as stated, namely, each point will describe an elliptical orbit with the centre of mass at one of the foci. These orbits will last forever for point bodies, so in this idealized case the long-term behaviour is known. The two-body problem becomes more complicated, firstly if one considers extended bodies such as stars instead of point bodies in Newtonian theory, and secondly if one applies the more accurate Einstein's theory to the case of two point bodies. Let us briefly consider the second complication. In Einstein's theory the problem of two point bodies going round each other has not been solved exactly. However, approximate calculations indicate that the bodies do not follow elliptical paths, but follow an approximately elliptical path in which the ellipse precesses, that is, the ellipse turns round. Also, the bodies slowly spiral inwards

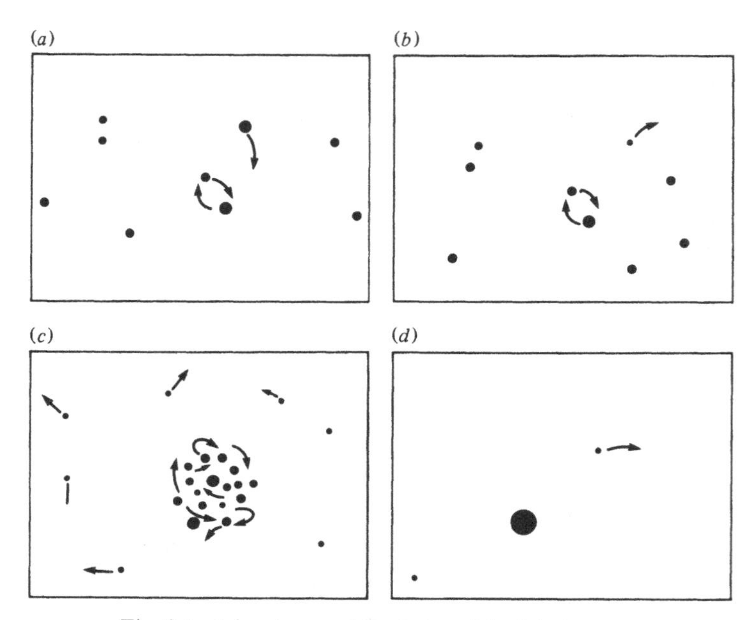

Fig. 8.1. Adventures of dead stars (black holes, neutron stars and white dwarfs) in a dying galaxy. In (*a*) and (*b*), a small 'star' orbiting another is dislodged and sent flying out of the galaxy by a more massive star, leaving the others tightly bound. After many such three-body, and more importantly, many-body interactions, the galaxy has a dense core (*c*), which finally collapses into a gigantic black hole (*d*).

towards each other while at the same time giving off a kind of radiation (not electromagnetic radiation) known as gravitational waves, which carries away energy from the system of two bodies. The two bodies thus become more tightly bound, that is, get confined to a smaller volume. (Although gravitational waves are predicted by Einstein's theory, they have not been detected yet in spite of strenuous efforts, in particular by J. Weber of the University of Maryland.) In practice one has extended bodies so that eventually the bodies may coalesce to form a single body unless some other process intervenes.

The second complication, namely, effects due to Einstein's theory, or General Relativistic effects, is quite important for compact bodies with strong gravitational fields such as two neutron stars. Such effects are very small when considering the Earth–Sun system or the Earth–Moon system. In fact there do exist two neutron stars which go round each other. This is the case of the binary pulsar PSR1913+16. J.H. Taylor and his collaborators at the University of Massachusetts are in the process of making a detailed analysis of the trajectories of the two pulsars. The indications are that this binary pulsar shows a number of general relativistic effects which are as predicted by Einstein's theory, including precession of the elliptical orbit and emission of gravitational waves. These results are preliminary and have to be confirmed by detailed analysis and more observations.

For two neutron stars going round each other general relativistic effects are quite important but for the Earth–Moon system, for example, other effects dominate. So as regards the first complication, namely two extended bodies in Newtonian theory, consider the case of the Earth–Moon system. It can be shown in Newton's theory that the gravitational field of a perfectly spherical body behaves outside the body as if its mass were concentrated at the centre of the body. This property in itself is not sufficient to reduce the problem of two spherical bodies to that of two point bodies, unless the bodies are perfectly rigid, because each spherical body feels the 'tidal' gravitational field of the other body which complicates the motion in the long run. In any case the Earth and the Moon are not perfectly spherical and not perfectly rigid. The Earth's gravitational field not only makes the Moon go round it, but it produces certain stresses in the body of the Moon itself. Likewise the gravitational field of the Moon produces stresses in the Earth, a well known example of which is tides in the ocean. These stresses produced in the Earth and the Moon affect their orbits round each other in the long run. In this system there is the additional complication that the Earth spins on its axis, as does the Moon (once in a lunar month). Because of the tides, there is a complicated way in which some of the angular momentum of the Earth is transferred to the Moon. As a result the length of the day is increasing (the rotation of the Earth is slowing down) by two milliseconds per century and the moon is spiralling away from the earth at about 3 cm/yr. Thus the spiralling in of two bodies that occurs due to gravitational radiation is utterly negligible for the short-term behaviour of the Earth–Moon system. Other effects, such as the transfer of angular momentum from the Earth to the Moon, dominate. However, in the very long run the Moon will stop receding when adequate angular momentum has been transferred. Then general relativistic effects may become important, but before this happens other processes will intervene such as the Sun's becoming a red giant, in which case both the Earth and the Moon will be swallowed up by the Sun.

To sum up the case of the two bodies, we know the orbits approximately but the very long-term behaviour depends on many details and is very difficult to predict exactly. The likelihood is that they will spiral into each other eventually to form a single body, unless some other external process intervenes. I have considered the two-body case in such detail partly to illustrate the fact that well-known solutions to simple problems sometimes become unsettled when we look into the far future.

Consider now a three-body system moving under mutual gravitational attraction. This problem cannot be solved exactly even for Newtonian theory and even when the bodies are considered to be point bodies. One has to solve the equations governing their motion numerically on the computer. However, one can give a qualitative analysis based on such numerical studies. As in the two-body case the three-body system will also emit gravitational waves and become more tightly bound. But before the emission of gravitational waves can have a significant effect on the orbits the likelihood is that one of the bodies will be ejected by coming into close collision with the other two. The ejected body may escape the system altogether while the remaining two come closer together. Thus the long-term behaviour of a three-body system is that one of the bodies will escape, leaving a two-body system. We have discussed the long-term behaviour of the latter.

As we introduce more and more bodies the problem becomes more complicated. For a typical galaxy after a thousand billion years or more we have essentially a hundred-billion-body problem. What will happen to this system eventually? As mentioned earlier, the exact behaviour of such a system is very difficult to predict, but some qualitative analysis can be made, which may turn out to be incorrect in some details. There are indications that such a system will form a dense central core of stars with a less-dense halo of stars around it (henceforth by 'stars' I mean dead stars, that is, cold white dwarfs, neutron stars or black holes). In the long run this configuration will change. There will be three-body encounters, and more frequently many-body encounters in which one of the bodies may acquire a high velocity. Most of the time the stars which will acquire high velocity will remain in the gravitation field of the galaxy, perhaps becoming a part of the extended halo. But occasionally after a many-body collision a star will acquire such a high velocity that it will escape from the galaxy altogether. Such three-body or many-body encounters are rare in time scales of billions of years. But in times of a billion billion (10^{18}) years or a million billion billion (10^{24}) years most of the stars in the galaxy will escape in this manner. The rest of the stars will form a very dense central core. As more and more stars escape, the remaining stars become more and more tightly bound, that is, they get confined to a smaller and smaller volume. This follows from consideration of the conservation of energy. As the central core becomes more and more dense, stars will coalesce to form larger stars, usually black holes, as white dwarfs and neutron stars cannot be too large. Ultimately the central core will coalesce to form a single large black hole. These close collisions during the formation of a single 'galactic' black hole will produce some fireworks, that is, light and other forms of radiation from the centre of the galaxy. There will be generation of energy near the centre in a manner similar to that in which quasars are thought to derive their energy. Namely, as the central black hole gets bigger, in the process of swallowing other dead stars the tidal gravitational field of the central black hole will disrupt these stars, which in turn will produce energy in the form of radiation. The number of stars which will escape and the number which will remain to form the galactic black hole is again very difficult to predict. This is one of the unsolved problems of astronomy. We can say tentatively that 99% of the stars will escape and 1% will remain so that the galactic black hole will have a mass of about a billion solar masses, with a Schwarzschild radius of about 3 billion km or 2–3 light hours. This is roughly half the distance from the Sun to Pluto, the furthest planet in the solar system.

A cluster of galaxies is likely to remain gravitationally bound as the expansion of the universe proceeds, that is, there will be no recession of galaxies within the cluster. Consider a cluster containing about a hundred galaxies. Each galaxy in the cluster will be reduced to a galactic black hole as above. In the very long run the cluster as a whole will evolve towards a single 'supergalactic' black hole (99% of the stars having escaped) with a mass of about a hundred billion (10^{11}) solar masses, having a Schwarzschild radius of about 300 billion km, or about a light week. The time scale for this will probably be anywhere from a billion billion (10^{18}) to a billion billion billion (10^{27}) years. The cluster of galaxies may in fact evolve to a single galaxy before the galactic black holes are formed. This will happen through the process of dynamical friction mentioned earlier by which process larger galaxies swallow up their smaller neighbours.

The process of the emission of gravitational waves, mentioned earlier, might make some contribution to the reduction of a galaxy to a single black hole. Calculations, however, indicate that this contribution will be negligible, as the dynamical effects described above will dominate before any significant changes can take place via gravitational radiation. The time scale for a galaxy to reduce to a single black hole through gravitational radiation is more like 10^{24}–10^{30} years. Thus this process will not be important in the reduction of a galaxy or a cluster of galaxies to a single galactic or supergalactic black hole.

In summary, in about 10^{27} years or so all galaxies and clusters of galaxies will have been reduced to galactic or supergalactic black holes. There will be a large number of dead stars and other pieces of matter (which were ejected from galaxies) wandering singly in the vast and ever-expanding emptiness of the intergalactic space. The galactic and supergalactic black holes will continue to recede from each other. Such time scales as I have been discussing in this chapter are likely to be available only if the universe is open, that is, if it expands forever.

9

A black hole is not forever

In the last chapter we saw that after a billion billion billion (10^{27}) years or so the universe will have two classes of black holes. Firstly there will be the very massive ones, namely galactic and supergalactic black holes. Another class of black holes will be the singly wandering stellar-size black holes (up to a few times the mass of the Sun) which were ejected from galaxies during the stage of dynamical evolution of the galaxy into a single black hole. There will, of course, also be the cold white dwarfs, neutron stars and other smaller pieces of matter (which were thrown out of galaxies) wandering in the intergalactic space. According to the laws of classical physics, all these black holes, white dwarfs, neutron stars etc. will last forever in the same form with very little further change. Perhaps we should explain here what we mean by 'classical' physics. 'Classical' here does not refer to classical Greece, but to a more modern period. Modern physics in one sense could be said to have started from the work of the Italian mathematician, astronomer and physicist Galileo Galilei (1564–1642) and of Newton. Nearly all the physical phenomena encountered in chemistry, physics and astronomy until about the end of the nineteenth century could be explained in accordance with the mechanistic principles propounded by Galileo and Newton. However, in the twentieth century it was realized that microscopic phenomena and also phenomena involving high velocities and strong gravitational fields could not be explained in terms of the laws of mechanics of Galileo and Newton. For these phenomena the Newtonian laws have to be replaced, respectively, by the laws of quantum mechanics and by the

Special and General Theories of Relativity. The physics that is based on the Galileo–Newton principles and which explained most phenomena until about the end of the nineteenth century is usually referred to as classical physics. The expression classical is used to contrast it with the more modern quantum physics or physics involving the Relativity Theory, either Special or General.

According to the laws of quantum mechanics, which supercede those of classical physics especially as regards microscopic phenomena, black holes, white dwarfs and neutron stars will suffer further changes. I shall deal with white dwarfs, neutron stars and other pieces of matter in the next chapter. In this chapter I shall discuss the long-term future of black holes as implied by the laws of quantum mechanics. It was discovered by S.W. Hawking of Cambridge University, that when quantum mechanics is taken into account black holes are not perfectly black but radiate in very minute amounts (for stellar size and bigger black holes). By this radiation the black holes lose mass and become smaller, eventually to disappear altogether in a final burst of radiation. This comes about as follows.

According to the laws of quantum mechanics, 'empty' space is not in fact perfectly empty but is full of *virtual* pairs of particles and antiparticles which are constantly being created and then destroyed. As mentioned earlier, for every kind of particle in nature there exists an 'antiparticle' with opposite charge but the same mass which has the property that a particle and its antiparticle can be created in pairs from pure energy or radiation. The pairs of particles and antiparticles filling empty space are called 'virtual' because unlike real particles, they cannot be detected directly with a particle detector. Their indirect effects, however, can be measured and their existence has been confirmed by a small shift (the Lamb shift) they produce in the spectrum of light from excited hydrogen atoms. The energy required to produce these virtual particles comes from the Uncertainty Principle (ennunciated by the German physicist Werner Karl Heisenberg (1901–1976)) which states, among other things, that if a system exists for a very short time, its energy is necessarily uncertain by an amount that depends on the duration of its existence. The shorter the duration, the more uncertain is the energy. It is as if the virtual particles, because of their extremely short duration, are able to borrow the energy for their existence from a bank run by the Uncertainty Principle. This phenomenon is known as 'vacuum fluctuation' (see Figs 9.1 and 9.2).

Consider now a black hole. There will be vacuum fluctuations just outside the Schwarzschild radius of the black hole, that is, just outside the horizon. In the presence of a black hole, one member of a pair of virtual particles may fall into the black hole, leaving the other member without a partner to annihilate. The forsaken particle or antiparticle may fall into the black hole after its partner, but it may also escape to the surrounding space, where it appears as radiation emitted by the black hole (see Fig. 9.3). But this radiation has energy, which must have come from somewhere. In other words, the virtual particle now becomes a real particle, so its energy can no longer come from

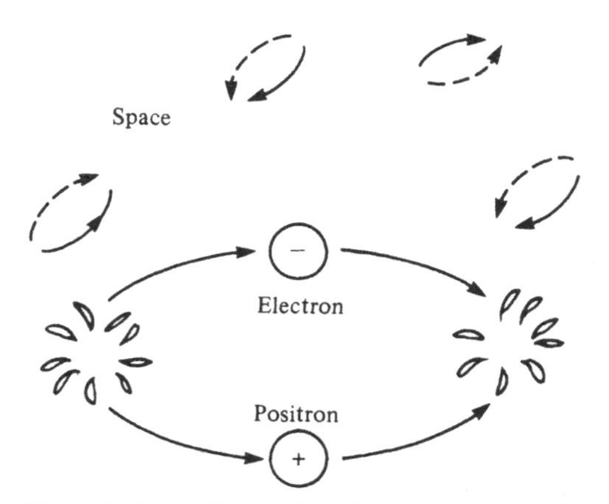

Fig. 9.1. According to the Uncertainty Principle in quantum mechanics, elementary particles like electrons can spontaneously appear in empty space with their antiparticles. (The antiparticle of an electron is an electron with positive charge, a positron.) These pairs of *virtual* particles, however, only exist on borrowed energy, and immediately merge and annihilate each other, leaving empty space again.

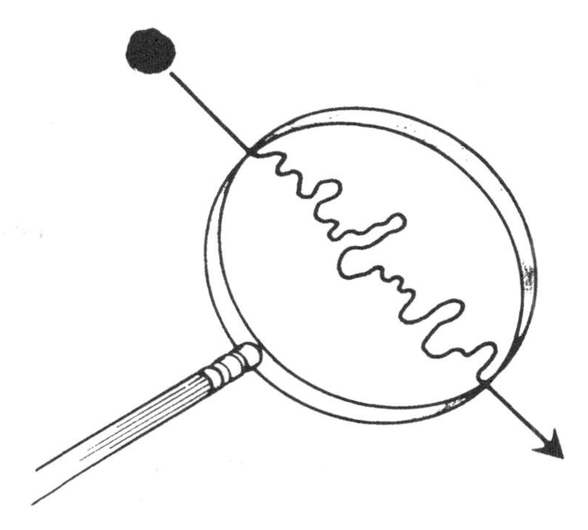

Fig. 9.2. Although the *virtual* particles depicted in Fig. 9.1 cannot be directly observed, their effect on real particles can be detected. Here an electron appears to be be travelling in a straight line, but on very small scales its path is curved and twisted erratically by the cumulative effect of tiny electrical fields produced by particles being continuously created and destroyed. Such a process produces a measurable shift in the energy levels of the hydrogen atom.

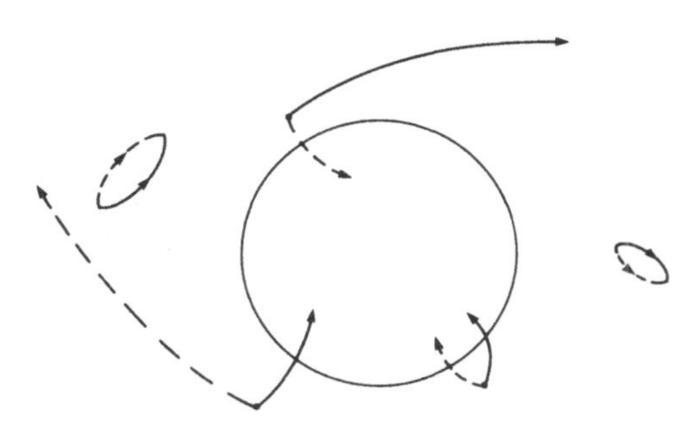

Fig. 9.3. In the neighbourhood of a black hole one member of a virtual particle–antiparticle pair may fall into the black hole, leaving the other member without a partner with which to annihilate. If the surviving member of the pair does not follow its partner into the black hole, it may escape to infinity. Thus the black hole will appear to be emitting particles and antiparticles.

'borrowed' energy due to the Uncertainty Principle. It can be shown that this energy in fact comes from the mass of the black hole. When one of the virtual particles goes into the black hole, it has negative energy with respect to an observer at a large distance. When this negative energy is added to the black hole, the black hole loses some of its mass, the energy corresponding to this reduction of mass appearing as the particle at a large distance, that is, as radiation from the black hole. In this manner the black hole radiates slowly and can be regarded as having a certain temperature. For a black hole of the size of the Sun the temperature is very low, namely, about a ten millionth of a degree absolute (10^{-7} K) and the radiation is in very small amounts. The black hole continuously loses mass as it radiates and eventually it evaporates completely in a final burst of radiation. The time scale for its eventual disappearance is about 10^{65} years for a black hole of one solar mass. The temperature of a black hole is inversely proportional to its mass. Thus the black hole has negative specific heat, that is, the more it radiates the hotter it becomes, unlike most ordinary bodies which become cooler when they radiate. The life-time of a black hole is directly proportional to the cube of its mass so that if the mass is doubled, the life-time becomes eight times as long.

A galactic black hole has a temperature of about 10^{-15} K whereas a supergalactic black hole has a temperature of about 10^{-18} K or a billion billionth of a degree absolute. Now as the universe expands, the temperature of the cosmic background radiation is inversely proportional to the scale factor or radius of the universe (denoted by R in Chapter 3). The background radiation will reach a temperature of 10^{-20} K in about 10^{40} years if we have a Friedmann model of the open universe with a 'flat' (Euclidean) geometry. On the other hand, if we have a model of the open universe which has a hyperbolic geometry, this temperature will be reached in about 10^{30} years. If the temperature of the cosmic background radiation is higher than that of the galactic or supergalactic black holes, the black holes will absorb more energy than they radiate. It is, however, clear from the above discussion and the time scales given that some time after the galactic and supergalactic black holes are formed, the black hole temperature will exceed the cosmic background temperature and the black holes will radiate more than they absorb. A galactic black hole will disappear altogether through this radiation in about 10^{90} years, whereas a supergalactic black hole will last for approximately 10^{100} years. The stellar-size black holes that were thrown out of galaxies would also evaporate long before 10^{100} years, since they would all be much smaller than supergalactic black holes. Thus in 10^{100} years or so all black holes in the universe will have disappeared and all galaxies as we know them today will have been completely dissolved. The universe will then consist of the singly wandering white dwarfs, neutron stars and smaller pieces of matter which were ejected from galaxies during the stage of dynamical evolution of the galaxies. The expansion of the universe will continue in the sense that the density of matter will keep decreasing, that is, the space between the singly wandering pieces of matter will continue to increase. Thus there will be an ever-growing emptiness which will contain minute and ever decreasing amounts of radiation with an ever-decreasing temperature, approaching inexorably the absolute zero of temperature but never quite reaching it.

10

Slow and subtle changes

In quantum mechanics it turns out that phenomena which are forbidden in classical physics (such as particles escaping from a black hole) have a small, but real chance of happening by a mechanism called *tunneling*, whereby a particle crosses a 'classical' barrier. By a classical barrier we mean one that would be a barrier if only the laws of classical physics operated. Thus an electron which does not have sufficient energy to surmount the barrier produced by an electrical field bounces off the barrier and cannot penetrate it according to the laws of classical physics, as shown in the upper sketch in Fig. 10.1. The wavelike properties of matter in quantum mechanics, however, give the electron a small chance of getting through (see lower sketch in Fig. 10.1). This phenomenon of tunneling is important in radioactive decay of a heavy nucleus such as a uranium or a radium nucleus and also in some processes in electronics. Since quantum effects are essentially microscopic effects, it is difficult to give an example of the phenomenon of tunneling in terms of every day happenings, but presently we shall try to explain radioactivity in such terms.

We shall see that the phenomenon of quantum tunneling causes some slow and subtle changes in the remaining pieces of matter after all the black holes have gone, or even before the black holes disappear. These processes would not be possible according to classical physics, since the latter implies that the matter in the form of white dwarfs, neutron stars and other smaller pieces of matter would stay in the same form forever.

Consider the long-term behaviour of any piece of matter, such as a rock or an asteroid. It will eventually cool to absolute

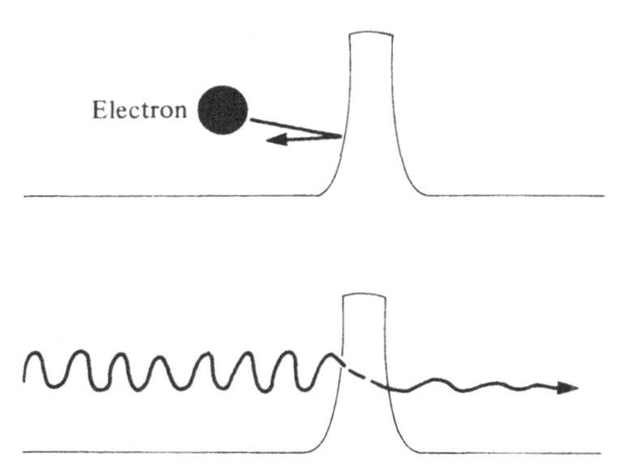

Fig. 10.1. In classical physics, electrons and other fundamental constituents of matter are particles. Thus, an electron which does not have sufficient energy to surmount the barrier produced by an electrical field bounces off as in the upper sketch. The wavelike properties of matter in quantum mechanics, however, give the electron a small chance of getting through. This phenomenon, *tunneling*, is important in electronics.

zero temperature (that is, 0 K). Its atoms will then be frozen into an apparently fixed arrangement by the forces of cohesion and chemical binding. According to classical physics, this arrangement will stay the same forever; but all matter seeks the state of lowest energy. There may be other arrangements of atoms which have a lower energy, but to get to these arrangements atoms may have to cross some electrical barriers which they cannot do classically. However, through quantum tunneling they can rearrange themselves to states of lower energy, but the time scale for this is very long. In fact even the most rigid materials will change their shapes and chemical structure on a time scale of 10^{65} years or so, and behave like a liquid, flowing into spherical shape under the influence of gravity. Suppose, for example, we constructed a small cube of the hardest diamond and managed to isolate it in space. Then in 10^{65} years or so the cube would become a sphere under the influence of gravity and because of the laws of quantum mechanics.

Radioactivity is a familiar phenomenon in nuclear physics. A typical example is when an alpha particle (I remind the reader that this is the nucleus of a helium atom, consisting of two protons and two neutrons) is ejected out of a heavy nucleus such as thorium or radium C′ (the latter is an isotope of the element radium; the atom of an isotope has the same number of electrons and protons as the atom of the original element, but it has a different number of neutrons. The chemical properties of an element and its isotopes are the same since these are determined by the number of electrons in the atom.) These heavy nuclei have a certain average life-time within which about half of the nuclei decay into lighter nuclei by giving off an alpha particle. The average life-time for thorium is 2×10^{10} years and for radium C′ it is about 10^{-3} s, that is, a thousandth of a second. This radioactive decay is not possible classically because the alpha particle is held to the nucleus by a certain force from which it cannot break loose according to the laws of classical physics due to an energy barrier. However, the alpha particle can 'tunnel' through the barrier and escape from the attractive force of the rest of the nucleus. This process, in which a heavy nucleus breaks up into lighter nuclei, is called 'fission' and is illustrated in Fig. 10.2. It is as if the particle exists in a well which is referred to in the figure as a 'potential well'. At any distance from the centre force exerted upon the particle is given by the slope of the potential well. Supposing the particle has energy E as shown in the figure. This energy is insufficient for it to go over the 'hump' and escape from the nucleus. At least this is the case as far as classical physics is concerned. But if the thickness of the barrier is finite at the energy level of the particle, it can 'tunnel' through this barrier and appear on the other side according to quantum mechanics. This kind of behaviour is completely inexplicable in terms of the everyday phenomena that we see around us and shows the manner in which the microscopic world differs from our macroscopic world.

There is also a nuclear reaction called 'fusion' in which two or more nuclei combine to form a heavier nucleus, such as the combination of four hydrogen atoms to form a helium atom, mentioned in Chapter 6. The latter process usually requires high temperature, but for certain nuclei which are in close

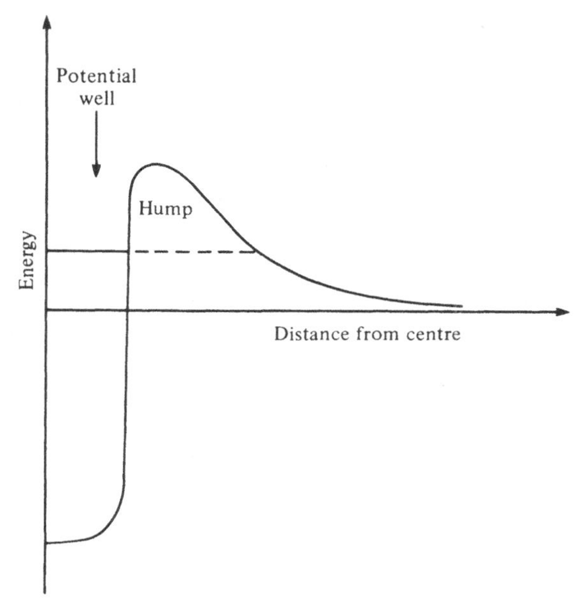

Fig. 10.2. Energy diagram to show the force felt by an alpha-particle in a heavy nucleus. For short distances (less than 10^{-13} cm) the alpha particle feels the attraction of the nuclear force while for larger distances (beyond the 'hump') it feels the repulsion of the electric force due to the other protons in the nucleus. Although the alpha particle does not have enough energy to go over the hump, it can 'tunnel' through the barrier.

proximity with one another this process can take place by quantum tunneling; again the time scale for this to happen by tunneling is very long. In fact in very long time scales *any* piece of ordinary matter becomes radioactive because it can release energy from nuclear fusion or fission reactions which take place by quantum tunneling. In the universe of the far future (after 10^{100} years) all pieces of matter other than neutron stars must decay ultimately to iron, which has the most stable nucleus. The time scale for this can be calculated by a formula given by G. Gamow. For the decay to iron the time scale is 10^{500}–10^{1500} years. On this time scale ordinary matter is radioactive and is constantly generating nuclear energy. Thus on this time scale our cube of diamond would become a sphere of iron.

What will eventually happen to white dwarfs and neutron stars? If a white dwarf is compressed from outside by some external agent, it will collapse to a neutron star. This is effectively what happens in a supernova explosion if the remnant is a neutron star. Due to this explosion of the outer layers, the central core gets compressed, and it temporarily reaches the white dwarf stage in which electrons are stripped off the atoms. As the compression continues, the electrons are squeezed into the protons to form neutrons and different nuclei merge together to form a neutron star. But consider the far future of the universe in which a white dwarf is wandering singly in space. In the near emptiness of the far future there will be no external agent to compress the white dwarf. However, the transformation of the white dwarf stage into the neutron star stage can be looked upon as the configuration of the white dwarf going 'over' a barrier and then going 'downward' to the neutron star stage, which is a stage of lower energy for the star as a whole. This barrier can be transcended spontaneously by the quantum mechanical tunnel effect even if there is no external agent doing the compression (see Fig. 10.3). The time

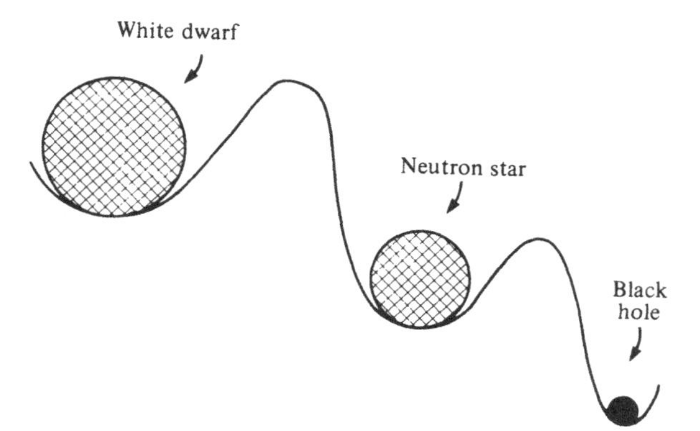

Fig. 10.3. Schematic diagram of the potential barriers that exist between the white-dwarf and the neutron-star states and between the neutron-star and black-hole states. These barriers can be crossed spontaneously by the tunneling effect.

scale for this can be calculated by the Gamow formula and is indeed very long. The time scale for this process (and indeed the time scales for other processes mentioned earlier in this chapter) has been calculated by F.J. Dyson of the Institute for Advanced Study at Princeton. The time scale for the collapse of a white dwarf to a neutron star through quantum tunneling is $10^{10^{76}}$ years!

The analysis of the last paragraph can be repeated for the collapse of a neutron star to a black hole. If a neutron star is compressed from the outside it will collapse to a black hole. This is the process by which black holes of mass less than $3M_\odot$ are created in supernova explosions. Again this can happen by quantum tunneling, the time scale being similar to that of the collapse of a white dwarf to a neutron star. Thus ultimately all white dwarfs and neutron stars will collapse spontaneously into black holes and eventually evaporate by the Hawking process.

Earlier in this book I discussed the formation of black holes whose masses are comparable to the mass of the Sun or are larger, such as galactic and supergalactic black holes. In principle it should be possible to have much smaller black holes than the mass of the Sun. For example, as mentioned earlier, the Schwarzschild radius for a mass of the Earth is about 1 cm. Thus if the earth were to be compressed to a radius of less than 1 cm, it would after that continue to collapse under its own gravity until it became a black hole. Unlike a large black hole (say ten times the mass of the Sun), which can be created by the force of gravity alone, a smaller black hole (such as one having a mass of that of the Earth or smaller) has to be created by compressing the matter artificially or by some other process. No known astrophysical process in the present universe can create small black holes, nor do we possess the technology to produce such black holes artificially. It was suggested by Hawking that such small black holes may have been created in the early universe by density fluctuations, that is, by the extreme variation in density from place to place perhaps due to turbulent and chaotic motion of the matter and radiation. Thus some small regions of space acquired a very high density, so that there was a great deal of matter in a very small region. If the region was smaller than the Schwarzschild radius of the matter contained in it, that mass would collapse and become a black hole. Whether such 'mini-black holes' exist in the present universe is not known. A mini-black hole of the mass of about 10^{15} g or less created in the early universe would have evaporated by now by the Hawking radiation process. Black holes of smaller mass than the Sun (including mini-black holes) are relevant when discussing the long-term stability of matter. Just as a neutron star may decay spontaneously into a black hole, so any piece of matter may collapse into a black hole (if black holes of the mass of that piece of matter are possible) by crossing a potential barrier by the tunnel effect. If an external agent compresses the piece of matter to within its Schwarzschild radius, it is effectively taking the piece of matter 'over' the potential barrier and 'downward' into the black hole configuration. But this can happen spontaneously without the external agent, causing the matter to decay to a black hole and evaporate subsequently by the Hawking process.

Consider now the long-term stability of matter as analysed by Dyson. The decay of white dwarfs and neutron stars may occur earlier than $10^{10^{76}}$ years if small black holes are possible. The minimum possible mass of a black hole is not known so there is some uncertainty in this analysis. Let M_B be the minimum size of a black hole, that is, suppose it is not in principle possible for a black hole to exist with mass less than M_B. Then the following alternatives arise.

(i) $M_B = 0$. In this case all matter is unstable with a comparatively short life-time. See Chapter 14 for a discussion of an experimental lower limit to the life-time of ordinary matter.

(ii) $M_B = 2 \times 10^{-5}$ g. This value of M_B, which is called the Planck mass, is suggested by Hawking's theory, according to which every black hole loses mass until it reaches the Planck mass at which point it disappears in a burst of radiation. In this case the life-time for all matter with mass greater than the Planck mass is $10^{10^{26}}$ years, while smaller pieces are absolutely stable.

(iii) $M_B = 3 \times 10^{14}$ g. This value of the mass is called the quantum mass because it is the mass of the smallest black hole for which in some sense a classical description is possible. In this case the life-time for a mass greater than the quantum mass is $10^{10^{52}}$ years, while smaller masses are absolutely stable.

(iv) $M_B = 2.8 \times 10^{33}$ g. This is called the Chandrasekhar mass because it is the maximum mass for a white dwarf. In this case the life-time for a mass greater than the Chandrasekhar mass is $10^{10^{76}}$ years, as discussed earlier.

The essential point in these alternatives is that all matter whose mass is greater than M_B can collapse spontaneously (without an external agent compressing it) into a black hole by the quantum tunnel effect and subsequently evaporate by the Hawking process. The long-term future of matter of all forms, and of the universe depends on which of the alternatives is correct. Dyson favours alternative (ii).

In the analysis so far we have been assuming that the 'stable' elementary particles such as electrons and protons are in fact absolutely stable. This may not be the case over periods which we have been discussing. This is an extremely important point and it will be discussed in detail in Chapter 14.

What, then, is the ultimate form of the universe if it is open? As is clear from the above discussion, the exact answer to this question depends on the long-term stability of matter. We will discuss in Chapter 14 the ultimate form of the universe if the proton is unstable. For the present, suppose that the minimum mass for a black hole is 2×10^{-5} g, that is, the Planck mass. Then in $10^{10^{26}}$ years all forms of matter whose mass is greater than the Planck mass will decay into radiation. This radiation will disperse and merge with the cosmic background radiation. The total density of matter and radiation will approach zero but never quite reach it. The density of matter, though approaching zero, will eventually (if not always) be greater than that of the radiation. The temperature of the background radiation will approach absolute zero but never quite reach it.

Will all physical processes cease eventually? As mentioned earlier, 'empty' space is not really empty but is seething with activity connected with vacuum fluctuations. Thus it looks as if there will be always microscopic phenomena in the emptiness of the far future universe even if all astronomical processes cease.

The concept of the passage of time loses some of its meaning when applied to the final stages of an open universe (or, for that matter, of a closed universe, as we shall see in Chapter 12). Time is measured against some constantly changing phenomena. If all astronomical processes cease, how will the passage of time manifest itself? It is doubtful if vacuum fluctuations can provide a clock for the recording of time. Will time itself come to a stop? Is this a meaningful question? Such questions are difficult to answer. I suppose eventually the only way in which the passage of time will manifest itself will be by the density and temperature of the cosmic background radiation, which will be decreasing all the time, approaching zero, but never quite reaching it.

11

Future of life and civilization

> I had a dream, which was not all a dream.
> The bright sun was extinguished, and the stars
> Did wander darkling in the eternal space,
> Rayless, and pathless, and the icy Earth
> Swung blind and blackening in the moonless air;
>
> *Darkness* by Lord Byron

It is almost impossible to predict what forms living organisms will take (assuming they can survive) in such time scales as we have been discussing. In an attempt to survive various extremely cold conditions, life may take forms which would be considered weird by our standards. However, the possibility of survival of life and civilization in any form depends on the availability of a source of energy, and one can discuss the latter. In this chapter I shall examine the sources of energy available, if any, during each of the stages of the universe described in the previous chapters. At each of these stages there will be enormous technical ingenuity required for civilization to survive. I will assume in the following that such technical ingenuity will be forthcoming. Very often civilization or society will have to face acute social problems. It might very well be that civilization may not survive some such problems, for example, a completely destructive nuclear war. I shall assume in the following that civilization will be able to achieve the maturity and wisdom to avoid such social catastrophies.

There will be adequate energy available as long as the Sun radiates sufficiently, which will be a few billion years. I do not necessarily mean here solar energy in the sense of the current discussion about sources of energy. Many forms of energy are indirectly dependent on the Sun. Take, for example, the energy from tides and winds. If the Sun were to become cold the oceans would freeze, as would eventually the air of the atmosphere, and there would be thus no wind and tides available. Similar remarks apply to any form of hydro-electric power, and many other sources of energy.

The Sun will most probably become a white dwarf, but before it gets to that stage it is likely to become a red giant (the Sun's mass is too low for it to undergo a supernova explosion). Either the Earth will be swallowed up in this process or in any event the Earth will become intolerably hot for living creatures. Civilization will then have to move either to one or more of the outer planets or reside in space colonies created artificially. The Sun's energy will still be available for use. In at most a few tens of billions of years the Sun will eventually become cold. By this time (in fact long before this time) civilization will undoubtedly be capable of surviving for millions of years on artificial nuclear energy. Civilization could then attempt to move to a different star, supporting itself during the transit by artificial nuclear energy. As the nearest star which is still radiating will probably be several light years away (the nearest star to us at present, Alpha Centauri, is about 4.34 light years away), the transit may take thousands of years or longer. It would be worthwhile making the transit because going in the neighbourhood of a radiating star may give a source of energy for several billions of years or several tens of billions of years.

In about a thousand billion years or some time afterwards, all stars in the galaxy will have been extinguished. It will not be worthwhile making the almost impossible attempt of going to a different galaxy (by this time the nearest galaxies will be much further apart than a million light years) since all galaxies will be in a similar dying state. There will, however, be another source of energy available within the galaxy, as explained in the following.

Most stars have a motion of rotation about an axis, much in the same way as the earth rotates once every day round its axis. There is some energy stored in this rotation and this is called rotational energy or energy associated with the angular momentum of the star. It is one of the well-verified laws of physics that angular momentum is conserved. The ice skater uses this principle when he increases his spin by drawing in his hands and leg near the axis round which he is spinning. Now when a rotating star finally collapses into a black hole, the rotation is preserved and serves to give the black hole rotational motion. Thus a rotating black hole has rotational energy which one can attempt to extract by means of a process suggested by R. Penrose of Oxford University. There is a region around a rotating black hole which is called the 'ergosphere' and which contains the horizon (the latter, readers will recall, is the region around a black hole out of which nothing can emerge). In the ergosphere it is impossible for a particle or observer to stand still since the rotation of the black hole forces the particle or the observer to revolve around the black hole. Further, it is possible for a particle in the ergosphere to have a negative energy with respect to observers outside and it is also possible for an observer or particle to leave the ergosphere and emerge outside (unlike the region within the horizon). In Penrose's process a particle with energy E_0 is sent to the ergosphere where it decays into two particles with energy E_1 and E_2, where E_1 is negative and E_2 is positive. The particle with energy E_1 goes into the black hole, while the one with energy E_2 comes out. The principle of conservation of energy says that we must have $E_0 = E_1 + E_2$. This equation implies that E_2 is greater than E_0, since E_1 is negative. Thus a positive amount of energy $E_2 - E_0$ is extracted from the black hole, this energy coming from the rotational energy of the black hole (see Fig. 11.1).

Civilization can find a rotating black hole in the galaxy and survive on the energy extracted from the black hole by the Penrose process. We are now in the stage of dynamical evolution of the galaxy (which will last about 10^{12}–10^{27} years) during which the galaxy is in the process of being reduced to a single galactic black hole. It is very likely that the parent star of the civilization (i.e. the rotating black hole from which energy is being extracted) will, sooner or later, be involved in a near three-body or many-body collision and hurled out of the

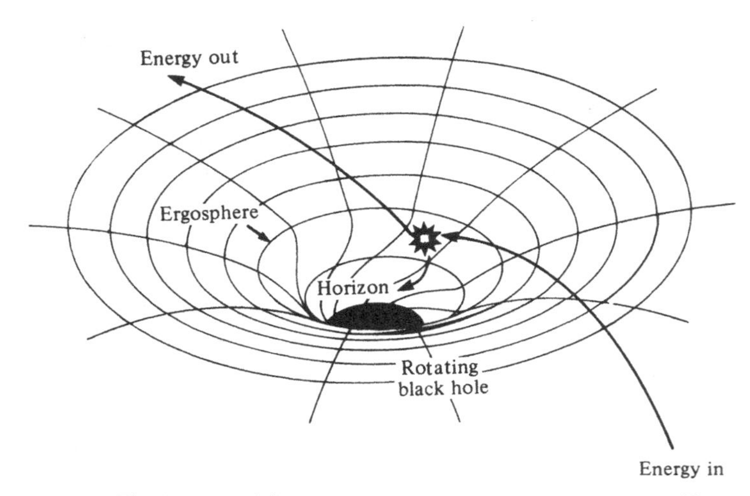

Fig. 11.1. Huddled around a supergalactic black hole 10^{27} years from now, future civilizations may depend on the mechanism discovered by Roger Penrose to extract energy from a rotating black hole. Such a hole is surrounded by a region called an ergosphere, into which it is possible to descend and come back out. A radioactive particle shot into the ergosphere at the right angle can decay so that one of its products falls into the black hole with negative energy; the other emerges from the ergosphere with more energy than the original particle, the extra energy having come from the black hole's rotation.

galaxy. It will then be opportune for the civilization to abandon the parent star and find another rotating black hole in the galaxy. I am assuming that societies will be residing in very mobile space colonies which can take appropriate action during near collisions etc. Those civilizations which survive until the galaxy reduces to a single black hole will probably crowd around the galactic black hole to extract its rotational energy. These societies may either be different splinters of our present civilization or they may be other intelligent societies from our Galaxy (if any exist). Since the Schwarzschild radius of the galactic black hole (strictly speaking a rotating black hole does not have a 'Schwarzschild radius' since the latter applies only to spherically symmetric black holes with no rotation, but this gives a rough idea of the size of the horizon of the rotating black hole) is about two or three light hours, it will be possible for different civilizations of the galaxy to communicate relatively easily with each other. As mentioned earlier, there will be numerous technical and social problems to solve all the time.

In principle this situation may continue for 10^{100} years, as long as galactic and supergalactic black holes exist and have rotational energy. Of course the rotational energy may be exhausted long before 10^{100} years. The Hawking radiation from galactic and supergalactic black holes will probably be too feeble to support a civilization.

It should be noted that the time scales we have been discussing are enormous compared with the present age of the universe, which is less than a mere wink of an eye compared with 10^{27} years or so for the formation of the supergalactic black hole. Thus evolutionary processes may take place which may be beyond our wildest imagination. In this we should note that the difference between 'natural' and 'artificial' becomes blurred. It could very well be that living creatures as we know them at present cannot survive indefinitely, but a new form of intelligent life created artificially especially adapted to endure in extremely cold surroundings may be able to survive indefinitely. This may not necessarily be an 'unnatural' process, for it may very well be essential for the indefinite survival of life to have as intermediary intelligent beings such as we are. Here of course one comes across the fundamental question, what does one mean by 'life'? I do not propose to go into this complex question here.

Thus after 10^{100} years or so, or long before that, external sources of energy will have been exhausted. The question whether civilization and life can survive indefinitely in an open universe boils down to the problem of surviving on a fixed finite amount of energy. It is very difficult to answer this question. Obviously conservation of energy will be very important. Even a minute rate of waste of energy, such as by radiation into the surrounding space, will amount to a substantial loss of energy over billions of years. It will be essential to overcome the problem of decay of matter, if matter is indeed unstable over long periods. In a fascinating paper (Dyson, 1979; cited at the end of this book) Dyson has raised and partially answered some fundamental questions about this problem. He asks if consciousness depends on the actual substance of a particular set of molecules or whether it depends only on the structure of the molecules. In other words, if a copy could be made of a brain with the same structure but using different materials, would the copy work as well as the original? If the answer is 'no', then life and consciousness can never evolve away from flesh and blood. Life can then continue to exist only so long as warm environments exist, with liquid water and a free supply of energy to support a constant rate of metabolism. In this case, since a galaxy has only a finite supply of free energy, the duration of life is finite.

Dyson, however, thinks that the basis of consciousness is 'structure' rather than 'matter'. In this case, he argues, a 'biological scaling law' may operate, in which as the temperature of the environment decreases by a certain factor, all the vital functions of the sentient creature are reduced in speed by the same factor. Dyson then gives arguments which make it plausible that, provided a society spends part of the time intermittently in hibernation with a reduced rate of metabolism (to conserve energy) the society can survive indefinitely on a fixed finite amount of energy. He also shows that such a society need not have a finite memory. In fact with the use of memory of the type of an analog computer it would in principle be possible for the society to have a memory of endlessly growing capacity. He also gives arguments to show that in principle it would be possible for different societies to communicate with each other over vast and ever expanding distances and times, with an expenditure of a finite amount of energy.

If Dyson is correct, he says, there is an analogy in physics and astronomy to the important theorem of Kurt Gödel in pure mathematics, which Gödel put forward in 1931, revolutionizing the subject of the foundations of mathematics. Gödel proved that the world of pure mathematics is inexhaustible; no finite set of axioms and rules of inference can ever encompass the whole of mathematics. Dyson says, 'If my view of the future is correct, it means that the world of physics and astronomy is also inexhaustible; no matter how far we go into the future, there will always be new things happening, new information coming in, new worlds to explore, a constantly expanding domain of life, consciousness and memory'.

12

A collapsing universe

In the previous chapters we have been concerned with the future of the universe if it is open, that is, if it will expand forever. The ultimate fate of the universe is dramatically different if the universe is closed, that is, if it will stop expanding at some future time and start to contract. If indeed the universe is closed, what is the time scale in which it will stop expanding and start to contract? This depends on the present average density of the universe. Models of closed universes can be constructed with arbitrarily long time scales for contraction by taking the present density to be above, but close enough to, the critical density mentioned in Chapter 5. Thus, in principle, it is possible to have a closed universe to expand for 10^{100} years before it starts to contract, so that most of the processes mentioned in the previous chapters will take place and then many of these processes will be reversed. If the universe is closed, however, it is extremely unlikely that its life-time will be as long as 10^{100} years.

Suppose for the sake of argument that the present average density of the universe is twice the critical density. Recall that in the simpler (Friedmann) models the closed universe has a finite radius. The universe will then expand until its radius is about twice its present value. The average distance between nearest galaxies, which is about a million light years now, will go up to about two million light years. The time required to reach the state of maximum expansion will be about 50 billion years. The temperature of the cosmic background radiation will go down to about 1.5 K and start to rise thereafter. There will not be much significant change noticeable in the universe during this time. After the turning point, all the major changes that took place in the universe since the big bang will be reversed. A few tens of billions of years after the maximum expansion the average density of galaxies will be the same as it is now, but instead of the red shift the distant galaxies will display a 'blue' shift (since the galaxies will be moving towards each other instead of away from each other; the blue shift will occur because visible light will be shifted in wavelength towards the blue end of the spectrum). A few billion years thereafter the temperature of the cosmic background radiation will rise to 300 K, and the sky everywhere, all the time, will be as warm as it is during the day at the present. After a few million years the temperature of the cosmic background radiation will rise to above 400 K and continue to rise thereafter so that the whole universe will be too hot for living creatures of any kind to survive. After some time the galaxies will merge with one another to form one continuous whole and soon afterwards stars will begin to collide at frequent intervals. It has been shown by M.J. Rees, that before the stars get disrupted by such collisions, they will in fact dissolve because of the intensity of the cosmic background radiation. When the latter reaches a temperature of about 4000 K, all electrons will be knocked out of atoms and finally, when the temperature reaches a few million degrees, all neutrons and protons will be torn apart from nuclei. Ultimately, there will be a universal collapse of all matter and radiation into a compact space of infinite or near infinite density in the so-called 'big crunch'.

Inevitably the question arises, what happens after the big crunch? This question is related to the one considered briefly in Chapter 3, namely what happened before the big bang? As in that case, no satisfactory answer exists to the question of what will happen after the big crunch. Indeed, it is not clear whether it is meaningful to talk about 'after' the big crunch, just as it is not clear whether it is meaningful to talk about what happened 'before' the big bang. These questions are not *necessarily* meaningless, but the fact is we simply do not know. One can make an attempt to analyse this question by asking about the nature of time and how the concept of time needs to be modified when we have the extreme conditions existing that prevail near the big bang or the big crunch. Such an analysis has been attempted by C.W. Misner of Maryland University. The point essentially is that all physical processes by which we measure time, such as the Earth going round the Sun, or an electron going round the nucleus of an atom, will all be gradually destroyed in the extreme conditions existing near the big crunch. One can therefore ask oneself what one means by the passage of time in such extreme conditions. Such an analysis can probably elucidate the problem somewhat, but it has not so far yielded an answer that is satisfactory and generally acceptable. It is clear that the phenomenon of gravitation dominates these extreme conditions. However, the most satisfactory theory of gravitation we have, namely Einstein's theory, may not be applicable in such extreme conditions, and one may have to modify it according to the laws of quantum mechanics. In other words, one needs a so-called quantum theory of gravitation. Some people, including Hawking, think that we may be able to understand the big bang or the big crunch (in particular, whether time has a beginning or an end at these events) when we have a satisfactory quantum theory of gravitation.

There is very little hope for life of any kind surviving the big crunch in a closed universe. However, one cannot be dogmatic about this as one does not know the limits of human ingenuity. If indeed the universe is closed, we probably have tens of billions of years to think about how to survive the big crunch, if it is not against the laws of nature that something should survive. One way in which a recurrence of life can occur is in the event that the cycle of the big bang and final collapse is repeated, and if galaxies are born again and again conditions for the existence of life may develop in some regions. Whether or not this can happen (assuming that the universe is closed) is, of course, not known.

13

The steady state theory

One of the most interesting of the non-standard models of the universe is the steady state theory, which has been the source of much controversy in the past. This controversy has, I believe, been healthy for the subject of cosmology, resulting in the creation of a great deal of interest in the subject and also stimulating new research which has led to important advances in astrophysics and cosmology. The steady state theory is currently not in favour for reasons which will be explained below.

The steady state theory was put forward by H. Bondi and T. Gold and independently by F. Hoyle in the same year (1948). The approach of Bondi and Gold was different from that of Hoyle, although the end result was the same. Bondi and Gold modified one of the cosmological assumptions to arrive at their theory, whereas Hoyle modified Einstein's equations.

In Chapter 3 I mentioned the Cosmological Principle, according to which the universe appears to be homogeneous and isotropic everywhere at any given time. The Principle of course allows the universe to evolve in time; in other words the universe can appear to be different at different epochs in its history. Bondi and Gold extended this principle to what is called the Perfect Cosmological Principle, according to which the universe is not only homogeneous and isotropic everywhere at any given time, but it appears on the average, to be the same at any time. Thus according to the Perfect Cosmological Principle, no large-scale changes take place in the universe as a whole. In particular, there is no 'big bang' in this model because the universe has always been the same as it is at present. One has, of course, to reconcile this model with the observed expansion of the universe. Now as the galaxies recede from each other, the average density of matter decreases. This is against the Perfect Cosmological Principle because by measuring the density we should be able to tell at which epoch of the universe we are. To compensate for this, the steady state theory postulates continuous creation of matter everywhere in the universe in very minute amounts by terrestrial standards. The amount of matter created to maintain a steady state depends on the present mean density of the universe and on the Hubble constant (see Chapter 3), but a plausible value is 4.5×10^{-45} kg/m^3/s. That is, for 1 kg of matter to appear in a cubical box with sides of length 1 m, we would have to wait 7×10^{36} years. Although this continuous creation of matter violates one of the most cherished laws of physics, namely the law of conservation of mass–energy, the amount by which the law is violated is so minute that it is not against any known experiments. We should also keep in mind that the problem of creation of matter also occurs in the standard model, for one can ask, where did the matter in the big bang come from? Thus the continuous creation of matter, although it violates the conservation of energy, is itself no reason to reject the steady state theory. The case against the steady state theory comes from observation.

In Hoyle's formulation of the steady state theory he avoids violating the conservation of mass–energy but at the expense of postulating a reservoir of negative energy in the universe. It can be shown that this reservoir of negative energy leads to continuous creation of matter and this rate of creation can be adjusted to yield a universe which is unchanging in time.

The main reason for discarding the steady state theory is the existence of the cosmic background radiation. As has been explained in Chapter 3 the existence of the cosmic background radiation implies that the universe has gone through a hot and dense early stage. This is against the steady state theory, which says that the universe should always present the same aspect in every epoch. There have been attempts to explain the cosmic background radiation within the framework of the steady state theory. For example, if it can be shown that the background radiation is not primordial but arises from sources existing in the past and present in an unchanging universe, then the background radiation would not imply a hot and dense early phase of the universe and so it would not be against the steady state theory. However, the attempts to explain the background radiation in this manner have not been successful.

Another piece of evidence against the steady state theory comes from quasars (discussed in Chapter 7). There are indications that quasars were more numerous in the past than they are at present. This indicates that the universe has evolved since the time when quasars were more in number and hence goes against a 'steady state' for the universe.

Although the universe does not change on average with time in the steady state theory, stars are born and die as usual as discussed in Chapter 6. As galaxies disperse, new galaxies are created in the intergalactic space by the condensation of newly created matter. Life and civilization could probably exist forever in a steady state universe as there would always be adequate energy available. The steady state theory also sidesteps the problem of the origin of the universe: the universe is as it is because that is the only way it can stay the same. The steady state theory is aesthetically and philosophically pleasing to many people, to whom it is a matter of regret that observations indicate that it is not the correct model.

14

The stability of the proton

In this chapter I shall consider one of the most important questions concerned with the long-term future of the universe and, indeed, one of the most important questions in physics. The question is whether or not the proton is stable. Until recently it had been assumed by physicists that the proton was indeed stable, that is, a proton left to itself would last forever. Recently, however, some theories of elementary particles have been put forward which imply that the proton is unstable, with a very long lifetime. In this chapter we shall try to see in what way these theories arise, and what are the consequences of proton decay. Before we can understand where the new theories fit, we shall have to know something about the theory of elementary particles, in much more detail than we considered in Chapter 4. To remind the reader I may repeat some of the points made earlier.

Every since the time of the ancient Greeks, people have wondered what is the ultimate nature of matter. They have wondered about the ultimate constituents of matter and about the manner in which these constituents affect each other or interact with one another. The Greek physical philosopher Democritus, who was born in the fifth century BC, speculated that all matter was made of atoms, which were eternal, indivisible and invisible. In the past hundred years or so and particularly in the last three or four decades a tremendous effort has gone into the investigation of this problem. The problem is to describe nature in terms of as few basic constituents as possible. The process of reducing the number of basic constituents and their interactions is still continuing. It seems nature can be described in terms of certain elementary particles and their mutual interactions. As mentioned earlier, all ordinary matter that we come across in everyday life is made out of electrons, protons and neutrons. These make up atoms which have a nucleus consisting of protons and neutrons and around the nucleus are electrons in orbits, the number of electrons in general being the same as the number of protons. Both the electron and the proton have electric charges, the electron having a negative charge while the proton has an exactly equal amount of positive charge. The neutron has no charge, that is it is neutral. All charged particles have the property that like-charged particles repel each other while unlike-charged particles attract each other. Thus all charged particles exert a force on each other and this force is referred to as the electromagnetic force. The 'magnetic' part arises from the fact that when charges are in motion they exert a different kind of force on each other than pure attraction or pure repulsion. This different kind of force is akin in some sense to magnetism. In fact all magnetic phenomena such as the attraction of magnetic iron for other pieces of iron can be shown to arise from charges in motion.

The nucleus of an atom, consisting of protons and neutrons, has a positive charge and it keeps the negatively-charged electron in orbit around it because of the electric attraction. The atom as a whole has no charge because there are equal numbers of electrons and protons. If an atom gains an electron it becomes negatively charged, or if it loses an electron it becomes positively charged, with one unit of charge in each case. In this case the atom is called an ion.

Soon after Rutherford's discovery that the nucleus of an atom is very much smaller than the atom itself and that it contains positively-charged protons, the question arose as to why the protons in the nucleus do not repel each other and cause the nucleus to disintegrate. The question was finally answered in 1935 by the Japanese physicist Hideki Yukawa (1907–1981) who suggested that there is a new kind of force which acts between a proton and a proton, between a proton and a neutron, and between a neutron and a neutron. The existence of this force has long been confirmed in numerous experiments. This force is referred to as 'nuclear force' or forces of 'strong interaction'. It is called 'strong' because it is much stronger (by a factor of about a hundred) than the electromagnetic force between charged particles, but it acts only over a very short range, namely, about 10^{-13} cm, which is the size of a nucleus. Beyond this range the nuclear force diminishes to almost zero whereas the electromagnetic force between charged particles diminishes only gradually with distance and has a long range.

What was the fact that finally convinced people that Yukawa was right? This fact was the discovery of an elementary particle predicted by Yukawa. On what grounds did Yukawa predict this particle? One of the facts that have emerged from the study of elementary particles is that these particles interact with each other through the exchange of elementary particles themselves. Thus an electron and a proton interact with each other by exchanging a particle of light or electromagnetic radiation, introduced earlier, called a photon. A simple example of this process is represented by Fig. 14.1*a*. In fact all charged particles interact with each other by exchanging photons. All phenomena involving electromagnetic forces can be shown to arise from such exchanges. As mentioned earlier, photons have zero mass, although they have energy. Because the photon has zero mass, the range of electromagnetic forces is infinite. That is, although these forces decrease with distance, they never go to zero, so that their range is infinite. In fact, electromagnetic forces decrease as the reciprocal of the distance. Roughly speaking, the range of a force is inversely proportional to the mass of the lightest particle that is being exchanged in the force concerned. We have seen that the range of the nuclear force is 10^{-13} cm. From this fact Yukawa was able to deduce that there must exist a particle in nature with mass about two hundred times the mass of the electron whose exchange causes the short-range nuclear force between a proton and a proton etc. (see Fig. 14.1*b*). In 1937 a new particle, the muon, was discovered with a mass 206.8 times the electron mass, but this did not fit in with other properties required of Yukawa's predicted particle. In fact Yukawa's prediction was confirmed later

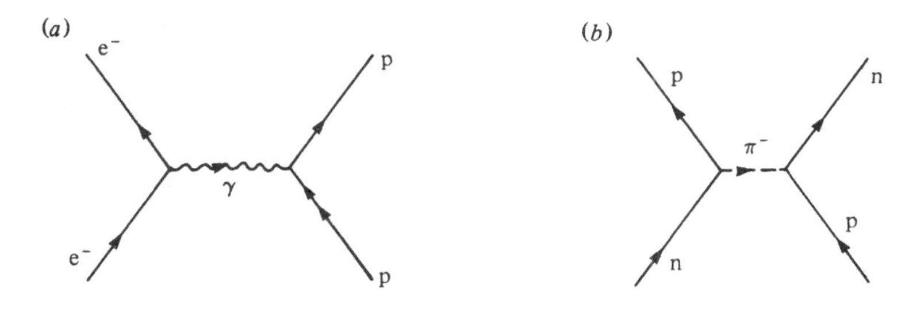

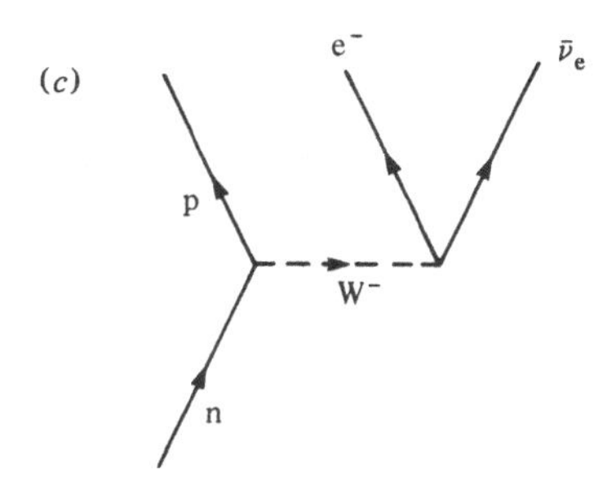

Fig. 14.1. This figure illustrates how forces are mediated by the exchange of particles. In (*a*) an electron (e^-) and a proton (p) interact by exchanging a photon (γ). In (*b*) a neutron becomes a proton by emitting a π^- meson, which is then absorbed by another proton which subsequently becomes a neutron. In (*c*) the beta decay of a neutron is caused by the emission of an intermediate vector meson W^- which decays into an electron and an electron-antineutrino.

in 1946 when three particles were discovered, which had the expected properties. These were the neutral pion π^0 (the π here is quite distinct from the π used earlier) with a mass of 264.1 electron masses, a positively charged pion, π^+, with a mass of 273.1 electron masses, and a negatively charged pion, π^-, with mass equal to that of π^+. It was this discovery that finally convinced people of the correctness of Yukawa's arguments, and established the existence of a new kind of force, the nuclear force. It is this force that holds the protons and neutrons together in a nucleus, overcoming the electrostatic repulsion of the protons.

In addition to the two kinds of forces that we have encountered so far, namely, the electromagnetic and the nuclear or strong force, there is another force which acts between elementary particles. This is the so-called weak force or forces of weak interaction. It is called weak because it is much weaker than electromagnetic or strong force; in fact it is about a hundred thousand times weaker than the strong force. Like the strong force the weak force is also of short range; its range is much shorter than that of of strong force, being only about 10^{-15} cm. An example of weak force is the beta decay of the neutron mentioned earlier. Recall that a neutron which is free, that is, not a part of a nucleus, decays in a few minutes to a proton, electron and an electron-antineutrino. If we denote the neutron by n, the proton by p, the electron by e^- (the minus denoting its negative charge; e^+ is the positively charged positron) and the electron-antineutrino by $\bar{\nu}_e$ (the bar on the top of ν_e denotes that it is an antiparticle; 'ν' is the Greek letter 'nu') the beta decay of a neutron can be written as

$$n \rightarrow p + e^- + \bar{\nu}_e$$

It may not appear that the decay of a particle can be caused by a 'force' but this is indeed so. One way of seeing this is to consider the following process, closely related to the one above

$$p + e^- \rightarrow n + \nu_e$$

Here an electron interacts with a proton to produce a neutron and an electron-neutrino. This process takes place due to the same weak force which causes beta decay. Such processes take place constantly in the centre of the Sun for example.

A whole host of elementary particles have been discovered in the last two or three decades. The more common ones are shown in Table 14.1. This table does not contain the tau-lepton mentioned earlier. Only those particles whose life-times are greater than about 10^{-17} s are included. *Leptons* take part in weak and electromagnetic interactions only but not in strong interactions. The *mesons* in the table are particles exchanged in strong interactions and they are all lighter than the proton. They all have spin zero and are bosons. The *baryons* are

Table 14.1. *In this table are listed the more common elementary particles. Particles and their antiparticles have the same mass, same life-time and opposite charges, so they are listed in the same line. A symbol with a bar over it denotes an antiparticle; thus $\bar{\nu}_\mu$ is the muon-antineutrino. Leptons take part in weak interactions but not in strong interactions. All hadrons take part in strong interactions; they are made up of mesons (which are bosons) and baryons (which are fermions). All hadrons also take part in weak interactions. Baryons other than the proton and the neutron are called hyperons*

Group	Particle	Symbol	Charge (in units of proton charge)	Mass (Mev)	Life-time (seconds)	Spin (in units of $\hbar$)
	Photon	γ	0	0	infinite	1
Leptons	Neutrino	ν_e, $\bar{\nu}_e$	0	less than 0.0001	infinite (?)	$\frac{1}{2}$
Leptons		ν_μ, $\bar{\nu}_\mu$	0	less than 0.0001	infinite (?)	$\frac{1}{2}$
Leptons	Electron	$e^\pm$	± 1	0.51	infinite	$\frac{1}{2}$
Leptons	Muon	$\mu^\pm$	± 1	105.66	2.2×10^{-6}	$\frac{1}{2}$
Hadrons – Mesons	Pion	$\pi^\pm$	± 1	139.57	2.6×10^{-8}	0
Hadrons – Mesons		π^0	0	134.97	0.84×10^{-16}	0
Hadrons – Mesons	Kaon	$\kappa^\pm$	± 1	493.71	1.24×10^{-8}	0
Hadrons – Mesons		κ^0, $\bar{\kappa}^0$	0	497.71	0.88×10^{-10}	0
Hadrons – Mesons	Eta	η	0	548.8	2.50×10^{-17}	0
Hadrons – Baryons	Proton	p, $\bar{p}$	± 1	938.26	infinite (?)	$\frac{1}{2}$
Hadrons – Baryons	Neutron	n, $\bar{n}$	0	939.55	918	$\frac{1}{2}$
Hadrons – Baryons	Lambda hyperon	Λ, $\bar{\Lambda}$	0	1115.59	2.52×10^{-10}	$\frac{1}{2}$
Hadrons – Baryons	Sigma hyperon	Σ^+, $\bar{\Sigma}^+$	± 1	1189.42	8.00×10^{-11}	$\frac{1}{2}$
Hadrons – Baryons	Sigma hyperon	Σ^0, $\bar{\Sigma}^0$	0	1192.48	less than 10^{-14}	$\frac{1}{2}$
Hadrons – Baryons	Sigma hyperon	Σ^-, $\bar{\Sigma}^-$	± 1	1197.34	1.48×10^{-10}	$\frac{1}{2}$
Hadrons – Baryons	Cascade hyperon	Ξ^0, $\bar{\Xi}^0$	0	1314.7	2.98×10^{-10}	$\frac{1}{2}$
Hadrons – Baryons	Cascade hyperon	Ξ^-, $\bar{\Xi}^-$	± 1	1321.3	1.67×10^{-10}	$\frac{1}{2}$
Hadrons – Baryons	Omega hyperon	Ω^-, $\bar{\Omega}^-$	± 1	1672	1.3×10^{-10}	$\frac{3}{2}$

Table 14.2. *This table shows the conservation of charge, baryon number and lepton number before and after the beta decay of a neutron*

	n	→	p	+	e^-	+	$\bar{\nu}_e$
Charge	0		1		−1		0
Baryon number	1		1		0		0
Lepton number	0		0		1		−1

particles of half-odd-integer spins which are fermions and they all take part in strong interactions. *Hadrons* are particles which take part in strong interactions, that is, the mesons and baryons together make up the hadrons. All hadrons take part in weak interactions also. All baryons other than the proton and the neutron are called *hyperons.* It is possible to assign a 'lepton number' and a 'baryon number' to each particle (and its antiparticle) such that in any process, the total lepton number and baryon number before the process is the same as those after the process has taken place. Thus lepton number and baryon number are 'conserved'. An example of this is shown in Table 14.2, which shows the conservation of lepton number and baryon number in beta decay. The lepton or baryon number of an antiparticle is the opposite of (that is, negative of) the lepton or baryon number of the corresponding particle. Thus for example the proton has baryon number 1, whereas the antiproton has baryon number −1.

We have encountered three kinds of forces through which elementary particles interact with each other, namely, the electromagnetic force, the strong or nuclear force and thirdly, the weak force. There is a fourth force in nature through which particles interact with each other and this is the familiar gravitational force which keeps us Earth-bound and makes the Earth go around the Sun. However, the gravitational force makes almost no contribution to the microscopic structure and processes of ordinary matter because it is extremely weak. In fact the gravitational force between a proton and a proton is

Table 14.3. *This table gives some properties of the four kinds of forces encountered in nature so far, namely gravitational, electromagnetic, strong (nuclear) and weak forces. 'Particles exchanged' means the particles through the exchange of which the corresponding force is mediated. The 'graviton' is a hypothetical particle through the exchange of which gravitational forces are mediated*

	Gravitational force	Electromagnetic force	Strong (nuclear) force	Weak force
Range	infinite	infinite	10^{-13} to 10^{-14} cm	less than 10^{-14} cm
Examples	astronomical forces	atomic forces	nuclear forces	beta decay of neutron
Strength	10^{-39}	$\frac{1}{137}$	1	10^{-5}
Particles acted upon	everything	charged particles	hadrons	hadrons and leptons
Particles exchanged	gravitons (?)	photons	hadrons	intermediate vector bosons (?)

10^{-36} times weaker than the electrostatic force. However, unlike the strong or weak force, gravitation is of long range and unlike electromagnetic forces gravitational force is always attractive. This is why the gravitational force accumulates and dominates all astronomical processes. Only these four kinds of forces have been encountered in nature so far by physicists. In Table 14.3 we display some properties of these four kinds of forces.

Note that all particles other than the photon, neutrinos, electron and the proton are unstable. These other particles decay either via the weak interaction or the electromagnetic interaction into lighter particles. Thus the neutron, when left to itself, decays into the stable particles proton, electron and an electron-antineutrino. The proton is stable, that is it does not decay into lighter particles, because it is the lightest baryon and baryon number is conserved. A particle, of course, cannot decay into a heavier particle because then mass–energy (mass and energy are interconvertible) would not be conserved. Likewise the electron is stable because it is the lightest charged particle and charge is conserved.

In addition to the particles listed in Table 14.1 there exist numerous other 'particles' which are unstable against strong interaction, that is, they decay via the strong interactions in a time scale of the order of 10^{-23} s (this is the time that light takes to cross a nucleus). These particles leave no tracks on photographic plates as some other longer-living particles do, because of their short duration. Their existence can nevertheless be deduced from the interactions of other particles. These particles which last for 10^{-23} s or so are often called 'resonances'.

The basic theory which describes the elementary particles and their interactions is called *quantum field theory*. This theory is a synthesis of quantum mechanics and relativity. Although this theory has had many successes, there still remain fundamental difficulties. This theory was applied to the electromagnetic interactions independently by the American physicists R.P. Feynman and J. Schwinger and the Japanese physicist S. Tomonaga in the 1940s (this question was considered by Heisenberg and others in the 1930s). It was Dyson who showed that the theories of Feynman, Schwinger and Tomonaga, known as quantum electrodynamics, were equivalent. When people first tried to calculate certain rates of interactions of charged particles, instead of getting a small number which they should have got they got an infinite result. Later a mathematical procedure was developed (Dyson played an important part in this development) known as *renormalization* by which the infinities occurring in the calculation could be handled and finite sensible numbers extracted which could be tested against experiments. Although mathematically very unsatisfactory, this procedure led to excellent agreement with experiment. Take for example the property of the electron known as its magnetic moment. The latter, roughly speaking, arises from the fact that because of its spin the electron can be considered as a tiny bit of circulating current and all circulating currents produce a magnetic field. The magnetism thus produced by an electron is related to its magnetic moment. Quantum electrodynamics predicts for the value of the magnetic moment of the electron, in suitable units, 1.0011596553 whereas the observed value is 1.0011596577; these values are in remarkable agreement.

A unified theory of the electromagnetic forces and the forces of weak interactions was formulated independently by A. Salam in 1968 and by S. Weinberg in 1967 following some earlier work by S.L. Glashow. This unified theory belongs to a type of quantum field theory known as *gauge theory*, an example of which was formulated by R.W. Mills and C.N. Yang in 1954. Important elements in the Glashow–Salam–Weinberg theory were provided by P.W. Higgs, G.'t Hooft and T.W.B. Kibble.

Just as electromagnetic interactions between charged particles are mediated by the photon (having a spin of one unit of $\hbar$), so in the unified Glashow–Salam–Weinberg theory weak interactions are mediated by three particles W^+, W^-, Z^0 (also having spin of $\hbar$). The W^+ particle has one unit of positive charge, W^- one unit of negative charge and the Z^0 is neutral. These particles are known as *intermediate vector bosons*. They are bosons because they have integer spin, and the term 'vector' implies that this spin is one unit of $\hbar$. The W^+ and W^- particles had been conjectured a couple of decades ago but the Z^0 particle was introduced by Salam and Weinberg to make the theory 'renormalizable' so that one got sensible results in calculation instead of infinities. The introduction of the Z^0 particle implies new weak interaction processes between some particles and these processes have been observed. The W^+, W^-, Z^0 particles themselves have so far not been observed because due to the very short range of weak interactions they turn out to be very massive. The W particles must be at least as heavy as about 40 protons and the Z^0 particle must be at least as heavy as 80 protons. Experiments are now under way in an attempt to produce these particles. In Fig. 14.1*c* is shown how beta decay of the neutron is effected through an intermediate W^- particle. How does the neutron get the energy to produce such a heavy W^- particle? In the process shown in Fig. 14.1*c* the W^- particle is 'virtual' (not real) and exists for a very short time so that it is able to borrow the energy (which is equivalent to mass) for its existence from the Uncertainty Principle.

Thus a satisfactory unified theory of electromagnetic forces and the forces of weak interaction has been found. What about the strong interactions? No satisfactory theory for the latter exists although a possible theory has been emerging in the last few years. This theory is called quantum chromodynamics. An important ingredient of this theory is the conjecture that all hadrons (baryons and mesons) are made of more fundamental entities called 'quarks'. Quarks were introduced into the theory by M. Gell-Mann and independently by G. Zweig. Quarks exist in five 'flavours' called 'up', 'down', 'strange', 'charmed' and 'bottom'. A sixth predicted flavour is 'top'. These six flavours are denoted respectively by u, d, s, c, b, t. (The b and t are sometimes called 'beauty' and 'truth'.) Each quark is also supposed to come in three 'colours' (nothing to do with visual colours). The quarks have the unusual property that they have non-integral charge. Thus the u-quark has charge equal to two-thirds of the proton charge and the d-quark has charge equal to one-third of the electron charge. No individual quarks have ever been observed so far but there is strong indication that the baryons and mesons behave as if they were made of quarks and antiquarks (as usual corresponding to each quark there exists an antiquark). Within a baryon or a meson the quarks interact with each other by exchanging still other kinds of particles called 'gluons' of which there are eight kinds, depending on their 'colour' composition. The proton, for example, has the quark composition uud giving it a total electric charge of $\frac{2}{3}+\frac{2}{3}-\frac{1}{3}$ or $+1$ unit, whereas the neutron consists of the quarks udd with charges $\frac{2}{3}-\frac{1}{3}-\frac{1}{3}$ or zero. The pion π^+, for example, is made up of a u-quark and a $\bar{d}$-antiquark. The 'chromo' in quantum chromodynamics refers to the 'colour' properties of quarks and gluons. This is a complicated concept which I do not propose to go into.

The Glashow–Salam–Weinberg theory unifies electromagnetic and weak interactions. One way to look at this is as follows. At energies high compared to the masses of the

Table 14.4. *In one form of the grand unified theories there is a correspondence between leptons and quarks as shown in this table. See the text for the meaning of the quark symbols. The τ^- refers to the τ-lepton and ν_τ is the corresponding neutrino. Each of the quarks come in three 'colours' (nothing to do with visual colours)*

	Leptons		Quarks	
	Symbol	Charge	Symbol	Charge
First generation	ν_e	0	u	$+\frac{2}{3}$
	e^-	-1	d	$-\frac{1}{3}$
Second generation	ν_μ	0	c	$+\frac{2}{3}$
	μ^-	-1	s	$-\frac{1}{3}$
Third generation	ν_τ	0	t	$+\frac{2}{3}$
	τ^-	-1	b	$-\frac{1}{3}$

intermediate vector boson (say at 10^{12} ev) the masses of these bosons can be neglected so they behave effectively as spin one zero mass particles, that is, as photons. Thus at high energies the weak interactions behave like electromagnetic interactions, since the latter arise from photon exchanges. Recently there have been attempts to unify all three of the fundamental forces, namely, electromagnetic, weak and strong forces, into one unified theory. The conjecture is that at still higher energies (at 10^{24} ev), all three forces have the same strength and behave similarly. In one form of these theories, which are called theories of grand unification, there is a correspondence between leptons and quarks as shown in Table 14.4. Thus the electron (e^-) and the electron-neutrino (ν_e) correspond respectively to the u-quark and the d-quark and so on. At very high energies leptons and quarks behave similarly and a quark can be converted into a lepton (something which never happens at ordinary energies over ordinary periods). This means that the proton, which hitherto has been regarded as absolutely stable, may in fact have a small probability of decaying into a lepton, that is, may decay with a very long life-time. In one form of these grand unified theories initially put forward by H. Georgi and S.L. Glashow, the life-time of the proton is about 10^{31} years. This does not mean that one has to wait 10^{31} years to find out if this theory is correct. (Recall that the age of the universe since the big bang is only 10^{10} years.) The time 10^{31} is an average life-time and it means that if there is a collection of 10^{31} protons there should be a decay once a year. In about 1000 tons of matter there are about 5×10^{32} protons and neutrons (if the proton decays the neutron should also decay) and roughly 50 of them can be expected to decay each year.

Several groups of people are planning experiments on this scale. Smaller experiments are already under way in a gold mine in South Dakota, USA and 2300 m underground in the Kolar gold mines 100 km north of Bangalore in South India. Larger experiments are planned in a salt mine near Cleveland, in a silver mine in Utah and in an iron mine in Minnesota. There is also one in two tunnels under the Alps. These experiments have to be done deep underground to minimize the amount of cosmic rays passing through the material. Cosmic rays are high-energy particles with which the Earth is constantly bombarded from outside. A cosmic ray event may be mistaken for a proton decay and hence it is necessary to shield the material under observation from cosmic rays by placing it deep underground. The present experimental limit for the stability of the proton implies that the life-time of the proton is *at least* about 10^{29} years. Thus if indeed the proton life-time is 10^{31} years or even 10^{33} years or so, it should be possible to verify this fact experimentally within the next decade, and thus to determine whether the theories of grand unification are correct.

The theory of gravitation predicts that the proton should decay into leptons in a time scale of about 10^{45}–10^{50} years, even if grand unified theories are incorrect. This comes about as follows. The black hole is described completely by three properties or three parameters, namely, its mass, angular momentum, and its charge. The black hole has no memory of whether the matter that went into it consisted of baryons or antibaryons, leptons or antileptons. Thus baryon and lepton numbers are not conserved once baryons and leptons go into a black hole. Now just as quantum mechanics predicts that empty space is full of virtual particle–antiparticle pairs, so the quantum theory of gravitation predicts that empty space is also full of 'virtual' black holes of all sizes, including virtual mini-black holes. If a proton is left to itself long enough, one of these virtual mini-black holes may swallow the proton and cause it to decay into a positron because the black hole, even a virtual one, does not respect baryon numbers. Since charge is one of the parameters of the black hole, it cannot make the charge of the proton disappear so that we get a positron and perhaps some photons. The life-time for this decay is very long and has been calculated by Hawking. It is approximately 10^{45}–10^{50} years. Thus the quantum theory of gravitation requires that the proton will eventually decay, even if grand unified theories are not correct.

The instability of the proton has profound consequences for the long-term future of the universe. All ordinary forms of matter will be unstable with life-times of the order of the proton life-time. The cube of diamond mentioned earlier will disintegrate into electrons and positrons long before it has a chance to become spherical and to become a sphere of iron. Now, normally an electron and positron annihilate to produce photons or pure radiation. Will all electrons and positrons produced out of the decay of matter annihilate to produce a universe of pure radiation, becoming thinner and thinner as the universe continues to expand? This is not so, according to M.R. McKee and D.N. Page. They find that a substantial proportion of the electrons and positrons will never annihilate, at least in the model of the ever-expanding universe which has a Euclidean geometry. They find that in such a universe the radiation density will never greatly exceed the matter density (that is, density of electrons and positrons). In fact the ratio of the matter density to radiation density will approach the value 0.6. The galactic black holes will still last for 10^{100} years or so because the baryons that have gone into making up the black hole all lose their identity and are not affected by proton or neutron decay.

What about the long-term future of life? All forms of life

have protons and neutrons as essential constituents. How can they survive proton decay? Is it conceivable that intelligent life can devise some way of preventing proton decay, if indeed the proton is unstable? If not, then intelligent beings have eventually to construct life out of electrons and positrons, something which is not necessarily excluded by the laws of nature. As Dyson says, if it is indeed possible for life to survive after the decay of protons, the final condition of our descendants will be curiously similar to the situation described more than half a century ago by J.D. Bernal in his book *The World, the Flesh and the Devil:*

> One may picture, then, these beings, nuclearly resident, so to speak, in a relatively small set of mental units, each utilizing the bare minimum of energy, connected together by a complex of etherial intercommunication, and spreading themselves over immense areas and periods of time by means of inert sense organs which, like the field of their active operations, would be, in general, at a great distance from themselves. As the scene of life would be more the cold emptiness of space than the warm, dense atmosphere of the planets, the advantage of containing no organic material at all, so as to be independent of both of these conditions, would be increasingly felt. . . .
>
> Bit by bit the heritage of the direct line of mankind, the heritage of the original life emerging on the face of the world, would dwindle, and in the end disappear effectively, being preserved perhaps in some curious relic, while the new life which conserves none of the substance and all the spirit of the old would take its place and continue its development. Such a change would be as important as that in which life first appeared on the earth's surface and might be as gradual and imperceptible. Finally, consciousness itself may end or vanish in a humanity that has become completely etherialized, losing the close-knit organism, becoming masses of atoms in space communicating by radiation, and ultimately perhaps resolving itself entirely into light. That may be an end or a beginning, but from here it is out of sight.

And what if the universe is closed? How does proton decay affect such a universe? Supposing a new era begins after the big crunch, will the number of protons (or baryons) be the same in the next cycle? Will the protons retain a memory of their previous life in the earlier epoch of the universe when deciding to decay or not to decay? Will there be subsequent cycles of big bangs and big crunches? If so, will proton decay affect the cycles of the far future? There exist no answers to such questions at present.

15

Epilogue

In this book I have presented what can at best be a rough outline of the long-term future of the universe and its ultimate fate. A great deal more needs to be understood about this problem, as is clear from the preceding chapters. For example, what is the nature of the long-term stability of matter? If the universe is closed, what is the precise nature of the final collapse? Is it really possible for life and civilization to exist indefinitely in an open universe? Can intelligent beings survive indefinitely the social conflicts (all too familiar in our present civilization) that beset society? One of the most intriguing problems is to understand the precise nature of time, especially with regard to the big bang, the big crunch and the long-term future of an open universe. Formulating an exact definition of time is an old problem. The early Christian philosopher, Saint Augustine (354–430) gave a classic expression to this problem when he said, 'What then is time? If no one asks me, I know: if I wish to explain it to one that asketh, I know not.'

The study of the universe as a whole is a unique enterprise. At least in one sense one is seeking to understand the totality of things. We, as thinking beings, are as much a part of the universe as are neutron stars and white dwarfs and our destiny is inextricably bound up with that of the universe.

If the standard model is correct, the universe started in a state of high density and temperature, with all matter and radiation forming one great continuous mass. It is very remarkable that this undifferentiated soup should have the intrinsic property that in due course of time it develops into galaxies of which at least one creates life with such staggering complexity, subtlety and diversity and often such stunning beauty. It also creates thinking and feeling beings which in turn can contemplate the universe and study its properties and which can love and hate. The British mathematician and philosopher Bertrand Arthur William Russell (1872–1970) says 'A strange mystery it is that Nature, omnipotent but blind, in the revolutions of her secular hurrying through the abyss of space, has brought at last a child, subject still to her power, but gifted with sight, with knowledge of good and evil, with a capacity of judging all the works of his unthinking Mother'. It is irrelevant whether or not there are other forms of life in the Galaxy or in other galaxies. The fact that we are here provides an 'existence proof' as it is called in mathematics. To say that we are an accident of nature is to miss the point. The laws of nature are presumably eternal and immutable. They do not change in mid-stream and suddenly acquire the ability to create a pretty toy if circumstances arise. With the French philosopher and mathematician René Descartes (1596–1650) who said 'cogito ergo sum' ('I think therefore I am') we might say 'I exist, therefore I am a part of the laws of nature'.

The urge is irresistible to ask, are we an essential part of the plan and architecture of the universe? Is there a purpose to the universe? Of course one can immediately counter such questions by asking what one means by 'essential part' and 'purpose'. Perhaps such questions are improperly posed and should not be asked, but it cannot be denied that these questions arise in the mind. One is reminded of the Vienna-born Cambridge philosopher Ludwig Wittgenstein (1889–1951) who said: 'We feel that even when *all possible* scientific questions have been answered, the problems of life remain completely untouched. Of course there are then no questions left, and this itself is the answer.' One of the most intriguing things about the universe, which probably cannot be explained by scientific investigations, is that it exists and we, who are a part of the universe, are able to contemplate and study it. It is this existence that often creates a sense of wonder in the human mind that causes us to ask questions to which no answers are forthcoming. Wittgenstein said 'It is not *how* things are in the world that is mystical, but *that* it exists'.

It is perhaps worth noticing that we have arrived 'on the scene' at a fairly early date. By this I mean that the time scale that it has taken nature to create us is of the same order of magnitude as the age of the universe. The universe is about 10–15 billion years old, and the Earth about 4.5 billion years old. Life is supposed to have begun on Earth about 3 billion years ago. It would not have been possible to evolve life, because of the hostile conditions, in the first few billion years after the big bang. Thus we have been created almost as soon as the universe was in a position to create us. It is an interesting question how long the universe will continue to create entirely new forms of life, assuming that it is open. It is clear that when the universe is sufficiently cold it will not be possible for new forms of life to emerge.

It is possible that those strange sentient beings of the far-future cold universe will find contemplating a warm universe such as ours not very pleasant, much as a nocturnal creature shuns daylight. But the more speculative amongst them may look back to our universe and to the Earth as an ideal world full of sunshine and a supply of adequate energy to last for billions of years, a dream world which will have passed away never to return. And what do we human beings do with this ideal dream world of ours? We oppress each other, build nuclear weapons for each other's destruction, and plunder the resources of the Earth!

Different people have different attitudes to the question of the ultimate survival of human civilization and the possibility that at some future time all life and civilization may end. To some people it is not the fact that all life and civilization may eventually vanish that is regrettable, but the fact that life contains so much cruelty and suffering while it lasts. The human mind has a different attitude towards 'time' and 'space' as regards the survival of the human race. Doubtless there is a desire in human beings to exist everywhere in space, but there seems to be a much stronger desire to exist everywhere in time, or at least in future time.

It is possibly true that intelligent life with a sophisticated technology is needed for the eventual survival of life. Dinosaurs and many other species became extinct because they could not adapt themselves to changes in the environment. Of course many other species have lived through many crises. But it is doubtful that any species, other than human beings (or at any rate, intelligent beings) can survive, for example, the Sun's becoming a red giant and eventual cooling down of the Sun. Could the emergence of intelligent beings like us be one of nature's plans for the eventual survival of life through various extreme conditions?

It is clear that there is a very great deal to be learnt about the universe and the endless subtleties of its various manifestations. What about the moral side of man, or what people with a religious bent of mind would prefer to call the spiritual nature of man? How will this develop in the endless aeons of the future? Perhaps in most of these questions like Newton we are still standing on the shore while the great ocean of knowledge lies ahead. It is significant that after more than two centuries of the acquisition of knowledge eminent men of science still have similar feelings. Different people express this differently according to their beliefs or lack of them. A sort of Gödel's theorem may operate not only in physics and astronomy, as suggested by Dyson, but also in other fields, so that there may always be new worlds to explore in all branches of knowledge. Russell speaks of 'the inexhaustible mystery of existence'. The German-born Princeton mathematician and physicist Herman Weyl (1885–1955), who made important contributions to cosmology (through the so-called Weyl postulate, which in some sense is equivalent to the Cosmological Principle) said (Weyl's sense of the words 'open' and 'closed' are, of course, different from the sense in which these words are used in this book): 'Modern science, in so far as I am familiar with it through my own scientific work, mathematics and physics make the world appear more and more an open one, as a world not closed but pointing beyond itself . . . Science finds itself compelled, at once by the epistemological and physical and the constructive-mathematical aspect of its own methods and results, to recognise this situation. It remains to be added that science can do no more than show us this open horizon; we must not by including the transcendental sphere attempt to establish anew a closed (though more comprehensive) world.' A relevant statement from Heisenberg is the following: 'The scepticism against precise scientific concepts does not mean that there should be a definite limitation for the application of rational thinking. On the contrary, one may say that the human ability to understand may be in a certain sense unlimited. But the existing scientific concepts cover always only a very limited part of reality, and the other part that has not yet been understood is infinite.' Lastly, Einstein, perhaps the greatest revealer of the subtleties of nature since Newton, said: the world 'stands before us as a great eternal riddle'.

Glossary

Absolute luminosity The total amount of radiation emitted per unit time by any astronomical body. This indicates how intrinsically bright the source is.

Absorption line The series of dark lines in the spectrum of a luminous object are called absorption lines. These are caused by the absorption of light from the object by the surrounding medium at specific wavelengths depending on the material that is absorbing the light.

Angular momentum A quantity which represents the amount of rotatory motion in a body.

Andromeda nebula This is a large galaxy about 2 million light years away from our Galaxy. It is called M31 and also NGC 224.

Anisotropic Not having the same property in all directions.

Antiparticle A particle with the same mass and spin as another particle but with equal and opposite electric charge, baryon number, lepton number, etc. The antiparticle of the electron is the positron, the antiparticle of the proton is the antiproton and so on. Some neutral particles are their own antiparticles, such as the photon and the meson.

Apparent luminosity The total amount of radiation received per unit time per unit receiving area from an astronomical body. This quantity essentially describes how bright or faint the source seems to us on earth.

Baryons This is a class of particles which take part in strong interactions including the proton, neutron and some heavier particles called hyperons. *Baryon number* is the total number of baryons present in the system, minus the total number of antibaryons.

Beta decay The radioactive decay of a nucleus in which an electron is given off. The beta decay of a free neutron (that is a neutron not confined to a nucleus) happens in a few minutes when it decays into an electron, a proton and an electron-antineutrino.

Big bang An event which took place about 10–20 billion years ago when there was an explosion at every point of the universe and at which time all matter was at very high density and pressure.

Big crunch If the universe is closed, there will be a universal collapse of all matter in a fiery implosion. This event, which is the opposite of the 'big bang', is called the big crunch.

Black-body radiation This refers to radiation which is in equilibrium with matter in the sense that it absorbs and emits the same amount of energy in any wavelength. The energy density in any wavelength is the same as that of radiation emitted by a totally absorbing heated body.

Black hole Matter which has collapsed to an extremely small volume either under its own gravitational force (this happens when a star several times more massive than the sun eventually dies) or by compression under some external agent. Due to strong gravity no light can escape from inside a black hole.

Blue shift The shift of spectral lines (absorption or emission lines) towards shorter wavelengths, caused by the Doppler effect of an approaching source of radiation.

Bosons Particles which obey the Bose–Einstein statistics. These particles must have integral spins in units of $\hbar$. Examples of bosons are photons and all mesons.

Cepheid variables These are stars which vary in their brightness with a certain period. From a knowledge of their period, it is possible to determine how intrinsically bright they are, i.e. to find out their absolute luminosity. They are used as distance indicators of nearby galaxies.

Charm A property possessed by the charmed quark (c-quark) and all particles containing this quark.

Classical physics Physics based on Newtonian principles which does not use quantum mechanics or Special or General Theories of Relativity.

Closed universe A model of the universe in which the universe will stop expanding at some future time and start to contract, eventually collapsing in a universal implosion.

Colour A property which serves to differentiate three varieties of each type of *quark*. This has nothing to do with visual colour. It also differentiates eight varieties of *gluons*. All observed particles are 'colourless' or 'white' combinations of coloured quarks.

Cosmic background radiation Radiation that is found to be coming from nowhere in particular and to have a temperature of about 3 K. It is thought to be present everywhere in the universe and to be the remains of the hot radiation that existed in the early stages of the universe after the big bang.

Cosmic ray High-energy particles which enter the atmosphere from outer space.

Cosmology The study of the large-scale structure of the universe.

Cosmological term A term added by Einstein in 1917 to his gravitational field equations. Such a term would produce a repulsion at very large distances and would be needed in a static universe to balance the attraction due to gravitation.

Cosmological Principle The hypothesis that the universe is homogeneous and isotropic everywhere.

Critical density If the present average density of the universe is less than the critical density then it will expand forever, whereas if it is more than the critical density, it will stop expanding at some future time and begin to contract.

Deceleration parameter A parameter (a number) which measures by how much the expansion of the universe is slowing down.

Density The amount of mass per unit volume. Sometimes it refers to the amount of mass and energy per unit volume. In this case it is usually referred to as *mass–energy density*.

Deuterium An isotope of hydrogen. The deuterium nucleus contains one proton and one neutron.

Doppler effect Change in the frequency or wavelength of any signal (like light or sound) caused by relative motion of source and receiver.

Electrical charge Also called simply *charge*, it is an intrinsic property of all particles such as electron or proton. The electron has negative charge and the proton has an equal amount of opposite or positive charge. Electricity is just the flow of charged particles, the electrons.

Electromagnetic force This is the force which is experienced by charged particles, such as the electron and the proton, when interacting with each other.

Electron This is the lightest known charged particle. All chemical properties of atoms and molecules are determined by electrical interactions of electrons with each other and with atomic nuclei. Its mass is about 9×10^{-28} g.

Electron volt This is a unit of energy used in atomic physics. It is the energy acquired by an electron in passing through a voltage difference of one volt. It is equal to 1.6×10^{-12} ergs.

Elementary particle The basic constituents of all matter such as electrons, protons and neutrons are called elementary particles. This is a changing concept because now even the proton and neutron are thought to be made of quarks, but loosely speaking the particles in Table 14.1 are all referred to as elementary particles. The term also refers to 'resonances' or very short-lived 'particles' mentioned in the text.

Emission line A series of bright lines in the spectrum of a luminous object. These are caused by radiation from hot gas in the object. Certain emission lines correspond to radiation from certain materials.

Erg The unit of energy in the centimeter–gram–second system. The energy due to its motion of a mass of 1 g travelling at 1 cm/s is half an erg.

Euclidean space Space in which the postulates of Euclidean geometry hold. For example, the surface area of a sphere of radius r is $4\pi r^2$ and its volume is $\frac{4}{3}\pi r^3$.

Fermion A particle obeying Fermi–Dirac statistics. It must have half-odd integral spin. Examples are electrons, protons and neutrons. Fermions resist being compressed into a small volume.

Flavour Properties like 'charm' and 'strangeness' which are manifested by quarks and observed particles. It contrasts with *colour* which is manifested only by quarks and gluons.

Frequency The rate at which the crests of a wave pass any given point. The frequency is given by the speed of the wave divided by the wavelength.

Friedmann model The mathematical model of the universe based on the General Theory of Relativity and the Cosmological Principle (without the use of the cosmological term).

Galaxy A gravitationally-bound system of stars containing about 10^{11} stars. Our galaxy is called the Galaxy or the Milky Way.

Gauge theories A certain class of theories which is under intense study at present as a possible theory of electromagnetic, weak and strong interactions of elementary particles. The term 'gauge' means 'measure' and is used for historical reasons.

General Theory of Relativity The theory of gravitation put forward by Albert Einstein in 1915. It is a more accurate theory of gravitation than the older Newtonian theory but substantial differences between the two theories arise only in very strong gravitational fields such as near pulsars or black holes. Also called *General Relativity*.

Gluons Particles of zero mass through the exchange of which quarks interact with each other. They have not been observed so far but there are strong theoretical reasons for believing in their existence. There are eight varieties of gluons.

Gravitational waves Just as electromagnetic waves arise from interactions of charged particles, so gravitational waves arise from the gravitational interactions of all particles. Gravitational waves travel at the speed of light, about 300 000 km/s. They have not been observed yet but their existence is implied by General Relativity.

Graviton Hypothetical particle which mediates gravitational interactions. It is supposed to be massless (like the photon) and to have an intrinsic spin of two units of $\hbar$.

Hadron Any particle that participates in the strong interactions. Hadrons are divided into baryons (which are fermions) and mesons (which are bosons).

Helium The second lightest element. Its nucleus contains two protons and two neutrons. The helium nucleus is called an *alpha particle*. Helium has an isotope, namely, helium-three (whose nucleus contains two protons and one neutron).

Homogeneity This is the property of the universe according to which there are on the average the same number of galaxies in a given volume wherever the volume is located.

Horizon In cosmology 'horizon' refers to the distance beyond which no light signal would have yet had time to reach us. For the black hole 'horizon' means the surface out of which no light signals can emerge.

Hubble's Law The law which says that the velocity of recession of a galaxy is proportional to its distance (provided it is not too far or not too near). The *Hubble constant* is the ratio of the velocity to distance in this relation of proportionality.

Hydrogen The lightest element, whose atom consists of an electron and a proton.

Hyperbolic space Space in which Euclidean geometry is not satisfied. In particular the surface area of a sphere of radius r is more than $4\pi r^2$ and its volume is more than $\frac{4}{3}\pi r^3$.

Infra-red radiation Electromagnetic waves with wavelength between 0.0001 cm and 0.01 cm, intermediate between visible light and microwave radiation. Bodies at room temperature usually radiate mainly in the infra-red.

Intermediate vector bosons These are the three hypothetical particles which mediate the weak interactions. They are called W^+, W^- and Z^0. They have unit spin and are bosons. They have not been detected yet.

Interstellar medium The region between the stars which contains gas and dust and many chemical compounds.

Isotope An isotope of an atom is another atom whose nucleus contains the same number of protons as the original atom but a different number of neutrons. The chemical properties of an atom and its isotope are similar since these depend on the number of electrons, which is the same as the number of protons.

Isotropy This is the assumed property of the universe according to which it looks the same in every direction wherever the observer is situated.

Kelvin This is a scale of temperature which is like centigrade except that its zero is the absolute zero of temperature. The melting point of ice at a pressure of one atmosphere is 273.15 K (K for Kelvin).

Leptons A class of particles which do not participate in strong interactions, including the electron, the muon and the neutrino. *Lepton number* is the total number of leptons in the system minus the number of antileptons.

Light year The distance light travels in one year, equal to about 9.5 million million km.

Mean free path This is the average distance travelled by a given particle between collisions with the medium in which it moves. The *mean free time* is the average time between collisions.

Mesons A class of strongly-interacting particles including pions, kaons and others which have zero baryon number and are bosons.

Messier number The catalogue numbers of various nebulae and star clusters in Charles Messier's catalogue. Usually denoted by a prefix M; thus the Andromeda nebula is M31.

Microwave radiation Electromagnetic waves with wavelength between about 0.01 cm and 10 cm. Bodies with a temperature of a few degrees Kelvin radiate in the microwave band.

Milky Way The stretch of light across the sky which marks the plane of our galaxy. Sometimes our Galaxy itself is called the Milky Way.

Mini-black hole A black hole which is small compared to the black holes of the size of the Sun or greater. A black hole of mass 10^{15} g, for example, may be called a mini-black hole. According to Hawking such mini-black holes may have been created in the early universe by turbulent motion when the density of matter was very high.

Muon An elementary particle of negative charge, similar to the electron but 207 times heavier; like the electron it is a lepton. Denoted by μ.

Nebulae Astronomical objects with cloudlike appearance. Some nebulae are galaxies; others are actual clouds of dust and gas within our galaxy.

Neutrino A massless electrically-neutral particle having weak and gravitational interactions only. Denoted by ν. They come in at least two and possibly three varieties, known as the electron-neutrinos (ν_e), muon neutrinos (ν_μ) and the τ-neutrino (ν_τ). They have spin $\frac{1}{2}\hbar$ and are fermions.

Neutron An elementary particle with no electric charge which is about 1838 times heavier than the electron. It has spin $\frac{1}{2}\hbar$, and is a fermion and a baryon. It is a constituent of all atomic nuclei except that of hydrogen. It partakes of weak and strong interactions.

Neutron star An extremely dense star consisting mainly of neutrons.

Newton's constant Also called Newton's gravitational constant, it is the fundamental constant of Newton's and Einstein's theory of gravitation, denoted by G. In Newton's theory, the gravitational force between two bodies is G times the product of the masses divided by the square of the distance between them.

NGC catalogue This is the New General Catalogue of Dreyer. The NGC number of an astronomical source is the place occupied by that source in the catalogue. Thus the Andromeda nebula is NGC224.

Nucleosynthesis The process of the making of heavier nuclei from lighter nuclei at high temperatures is called *nucleosynthesis*. This occurred in the early universe and takes place in the centre of stars.

Nucleus The heavy central portion of an atom, consisting of neutrons and protons, is called the nucleus. Its size is about 10^{-13} cm compared with 10^{-8} for the atom as a whole.

Open universe The model of a universe which will expand forever in the future. In the simpler Friedmann models an open universe is infinite in spatial extent.

Parsec An astronomical unit of distance equal to about 3.26 light years.

Pauli Exclusion Principle The principle that no two particles of the same type can occupy precisely the same state. This principle is obeyed by fermions but not by bosons.

Perfect Cosmological Principle The Principle that the universe, on the average, appears to be the same everywhere and at all times. This Principle leads to the steady state theory.

Photon A particle associated with light waves or electromagnetic waves. Denoted by γ. It has no mass, spin $\hbar$, and is a boson. It mediates electromagnetic forces between charged particles.

Pion Also called π meson or pi-meson, it is the hadron of lowest mass. It comes in three varieties: positively charged (π^+), negatively charged (π^-) and neutral (π^0).

Planck's constant The fundamental constant of quantum mechanics denoted by h. It is equal to about 6.625×10^{-27} erg second. A related constant often used is $\hbar$ (h-slash) which is h divided by 2π (where π is a number approximately 22/7).

Positron The positively-charged antiparticle of the electron. Denoted by e^+.

Proton A positively-charged particle about 1836 times as massive as an electron. It has spin $\frac{1}{2}\hbar$, is a fermion and a baryon. It partakes of electromagnetic, weak and strong interactions.

Pulsars Stars discovered in 1967 which send out pulses at regular intervals of the order of a second. They are supposed to be rotating neutron stars.

Quantum chromodynamics This is the theory, still being worked out and incomplete, which describes strong interactions based on gauge theory of the colour properties of quarks and gluons.

Quantum electrodynamics The theory of the electromagnetic interactions between all charged particles and photons.

Quantum field theory The formalism which applies quantum mechanics to a *field*, that is a quantity which varies from point to point. It is a synthesis of quantum mechanics and special relativity.

Quantum mechanics A fundamental theory developed in the 1920s as a replacement of classical (Newtonian) mechanics to describe microscopic phenomena. In quantum mechanics, waves and particles are two aspects of the same underlying entity.

Quarks These are hypothetical fundamental particles of which hadrons are supposed to be composed. Isolated quarks have never been observed, but there are strong theoretical reasons for believing in their existence. Different types of quarks correspond to different 'flavours' and 'colours'. Quarks have charge either equal to two-thirds the proton or one-third the electron charge. Antiquarks have opposite charge and opposite values of some other attributes of quarks.

Quasars A class of astronomical objects with star-like appearance but which have large red shifts. They are thought to be galaxies which are far away in which violent events have taken place. However, there is controversy about their true nature.

Recombination This refers to the combination of atomic nuclei and electrons into ordinary atoms that took place in the early universe when the temperature dropped to about 3000 K.

Red giant A star of enormous size compared to the Sun but having a reddish colour because its surface is cooler than that of the Sun.

Red shift The shift of the spectral lines from a source towards longer wavelengths, caused by the Doppler effect from a receding source. From the red shift of a source it is possible to determine the velocity of recession of the source.

Renormalization The procedure whereby certain quantum field theories can be made to yield finite sensible results, instead of infinities, by identifying the total mass, charge, etc. with experimentally determined values.

Special Theory of Relativity This theory (also called *Special Relativity*) incorporates the new view of space and time as put forward by Albert Einstein in 1905. In this theory light (or electromagnetic waves) has the same velocity with respect to *all* observers, even if these observers have different velocities with respect to each other. This theory has the consequence that the rates of clocks of different observers who are moving with respect to each other are different, also that mass and energy are equivalent.

Speed of light The fundamental constant of Special Relativity, equal to about 300 000 km per second and denoted by c. Any particles of zero mass, such as photons, neutrinos and gravitons, travel at the speed of light.

Spherical space Space in which Euclidean geometry is not satisfied. In particular the surface area of a sphere of radius r is less than $4\pi r^2$ and its volume is less than $\frac{4}{3}\pi r^3$.

Spin A fundamental intrinsic property of elementary particles which describes the state of rotation of the particle. The spin is measured in units or half units of $\hbar$ (this is Planck's original constant h divided by 2π). Particles can have integral spins $0\hbar$, $\hbar$, $2\hbar$, . . ., etc. (these are called bosons) or half-odd-integral spins $\frac{1}{2}\hbar$, $\frac{3}{2}\hbar$, $\frac{5}{2}\hbar$ . . . (these are called fermions).

Steady state theory The cosmological theory developed by Bondi, Gold and Hoyle, in which the average properties of the universe are the same everywhere and never change with time. In this theory matter has to be created continuously in order to keep the density of the universe the same as it spreads.

Strong interaction The strongest of the fundamental forces through which elementary particles interact with each other. It is responsible for the force with which protons and neutrons are held together in an atomic nucleus. It affects all hadrons but not leptons or photons.

Supernova Enormous explosion of certain stars in which the outer portions of the star are blown off and the inner core compressed. In a supernova as much energy is produced in a few days as the Sun radiates in a billion years.

3C Catalogue This is the Third Cambridge Catalogue of astronomical radio sources prepared by the Mullard Radio Astronomy Observatory of the University of Cambridge.

Tritium An unstable isotope of hydrogen whose nucleus contains one proton and two neutrons.

Tunneling A microscopic phenomenon by which a particle crosses an electrical or other barrier by the laws of quantum mechanics. The particle would not cross the barrier if only the laws of classical physics operated.

Ultra-violet radiation Electromagnetic radiation with wavelength in the range 10^{-7} cm to 2×10^{-5} cm, intermediate between visible light and X-rays.

Uncertainty Principle According to this principle, enunciated by Heisenberg, one cannot determine precisely the position and velocity of a particle at the same time. Also, if a system exists for a short time, one cannot determine its energy precisely.

Vacuum fluctuations This refers to the constant or ever-present creation and annihilation of virtual particle pairs in empty space.

Virtual particles Particles exchanged during the interaction of real particles. Virtual particles cannot be observed directly but their indirect effects can be observed. The energy for their existence comes from the Uncertainty Principle.

Wavelength In any kind of wave, the distance between wave crests is called wavelength.

Weak interactions This is one of the four fundamental types of forces or interactions experienced by elementary particles. An example of this interaction is the beta decay of the neutron into an electron, a proton and an electron-antineutrino.

White dwarf A compact star about the mass of the sun but with the size of the earth. In the material of this star the electrons are stripped off the atoms due to strong pressure of gravitation and run around freely in the material of the star. The electrons provide the so-called 'Fermi pressure' which balances the force of gravity.

Selected bibliography

The references with an asterisk (*) are more technical than those without. The references without asterisks are suitable for the lay person. This bibliography is by no means complete.

Barrow, J.D. and Tipler, F.J. (1978) Eternity is unstable, *Nature*, **276,** 453 (*).

Davies, P.C.W. (1978) *The runaway universe*, J.M. Dent & Sons.

Davies, P.C.W. (1979) *The forces of nature*, Cambridge University Press.

Dyson, F.J. (1979) *Disturbing the universe*, Harper & Row.

Dyson, F.J. (1979) Time without end: physics and biology in an open universe, *Reviews of Modern Physics*, **51,** 447 (*).

Georgi, H. (1981) A unified theory of elementary particles and forces, *Scientific American*, **April**.

Gott, J.R., Gunn, J.E., Schramm, D.N. and Tinsley, B.M. (1976) Will the universe expand forever?, *Scientific American*, **March**.

Hawking, S.W. (1977) The quantum mechanics of black holes, *Scientific American*, **Jan**.

Hoyle, F. (1975) *Astronomy and Cosmology*, W.H. Freeman & Co.

Islam, J.N. (1977) Possible ultimate fate of the universe, *Quarterly Journal of the Royal Astronomical Society*, **18**, 3 (*).

Islam, J.N. (1979) The long-term future of the universe, *Vistas in Astronomy*, **23**, 265 (*).

Islam, J.N. (1979) The ultimate fate of the universe, *Sky & Telescope*, **Jan**.

Kaufmann, S. (1979) *Galaxies and quasars*, W.H. Freeman & Co.

Mitton, S. (1976) *Exploring the galaxies*, Faber & Faber Ltd.

Mitton, S. (1979) *The crab nebula*, Faber & Faber Ltd.

Narlikar, J.V. (1977) *The structure of the universe*, Oxford University Press (*).

Page, D.N. and McKee, M.R. (1981) Eternity matters, *Nature*, **291**, 44 (*).

Penrose, R. (1972) Black holes, *Scientific American*, **May**.

Polkinghorne, J.C. (1979) *The particle play*, W.H. Freeman & Co.

Rees, M.J. (1969) The collapse of the universe: an eschatological study, *The Observatory*, **89,** 193 (*).

Rees, M.J. (1981) Our universe and others, *Quarterly Journal of the Royal Astronomical Society*, **22,** 109.

Sciama, D.W. (1971) *Modern cosmology*, Cambridge University Press.

Thorne, K.S. (1967) Gravitational collapse, *Scientific American*, **Nov**.

Thorne, K.S. (1974) The search for black holes, *Scientific American*, **Dec.**

Wald, R.M. (1977) *Space, time and gravity*, The University of Chicago Press.

Weinberg, S. (1974) Unified theories of elementary-particle interaction, *Scientific American*, **July.**

Weinberg, S. (1977) *The first three minutes*, Andre Deutsch.

Index